AF412127

Environment and Object

Recent African Art

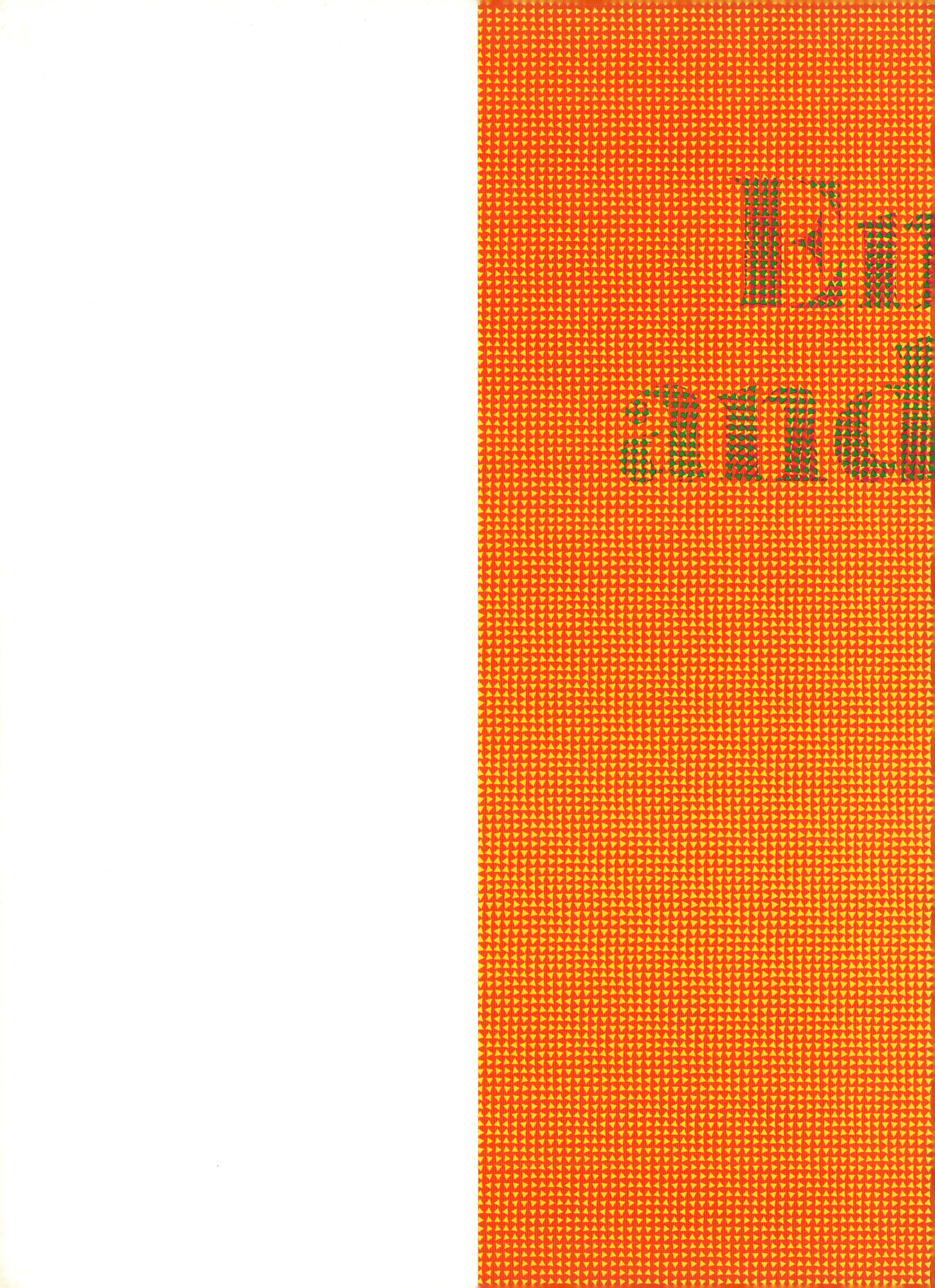

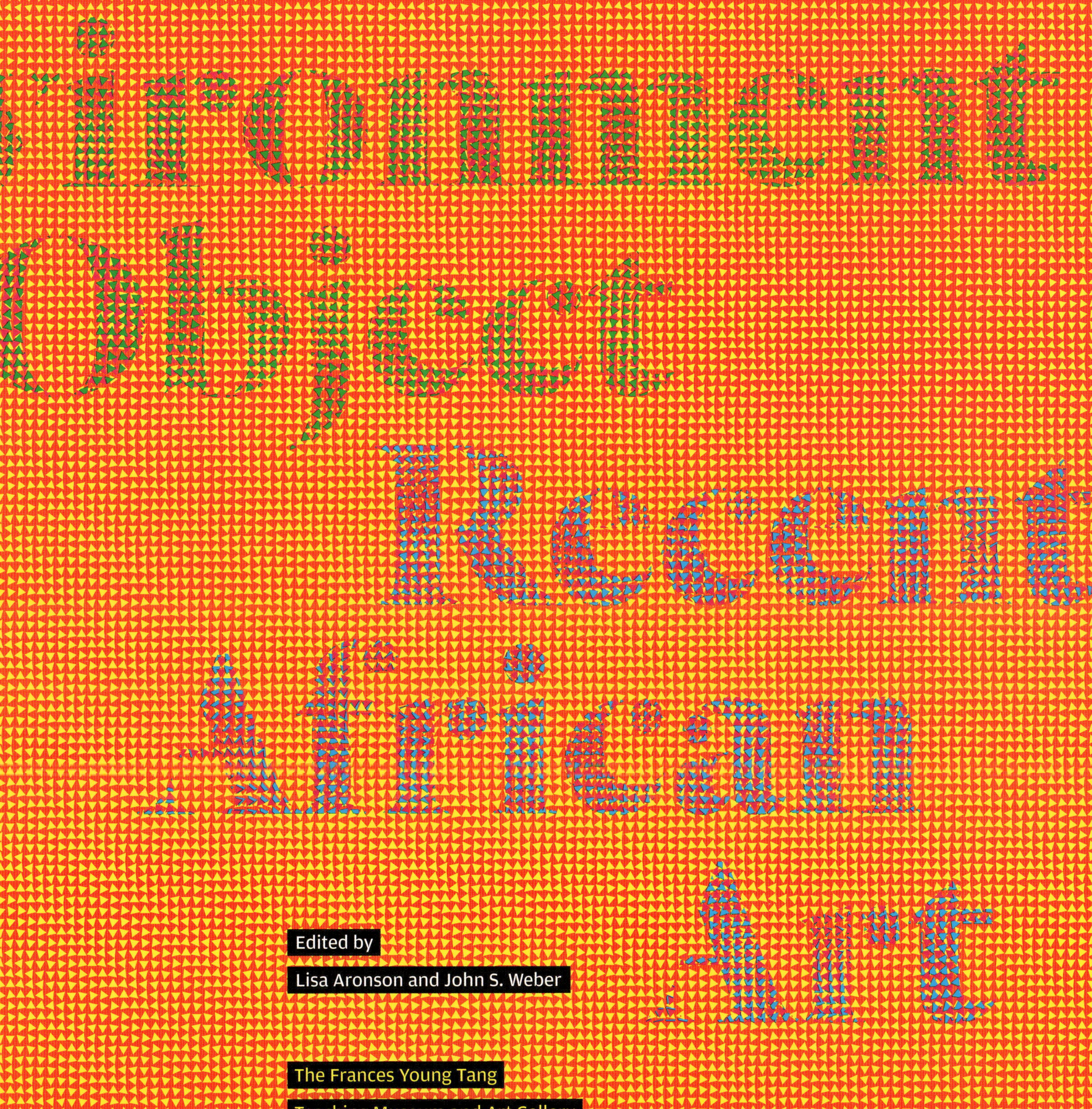

Edited by
Lisa Aronson and John S. Weber

The Frances Young Tang
Teaching Museum and Art Gallery
at Skidmore College

DelMonico Books • Prestel
Munich London New York

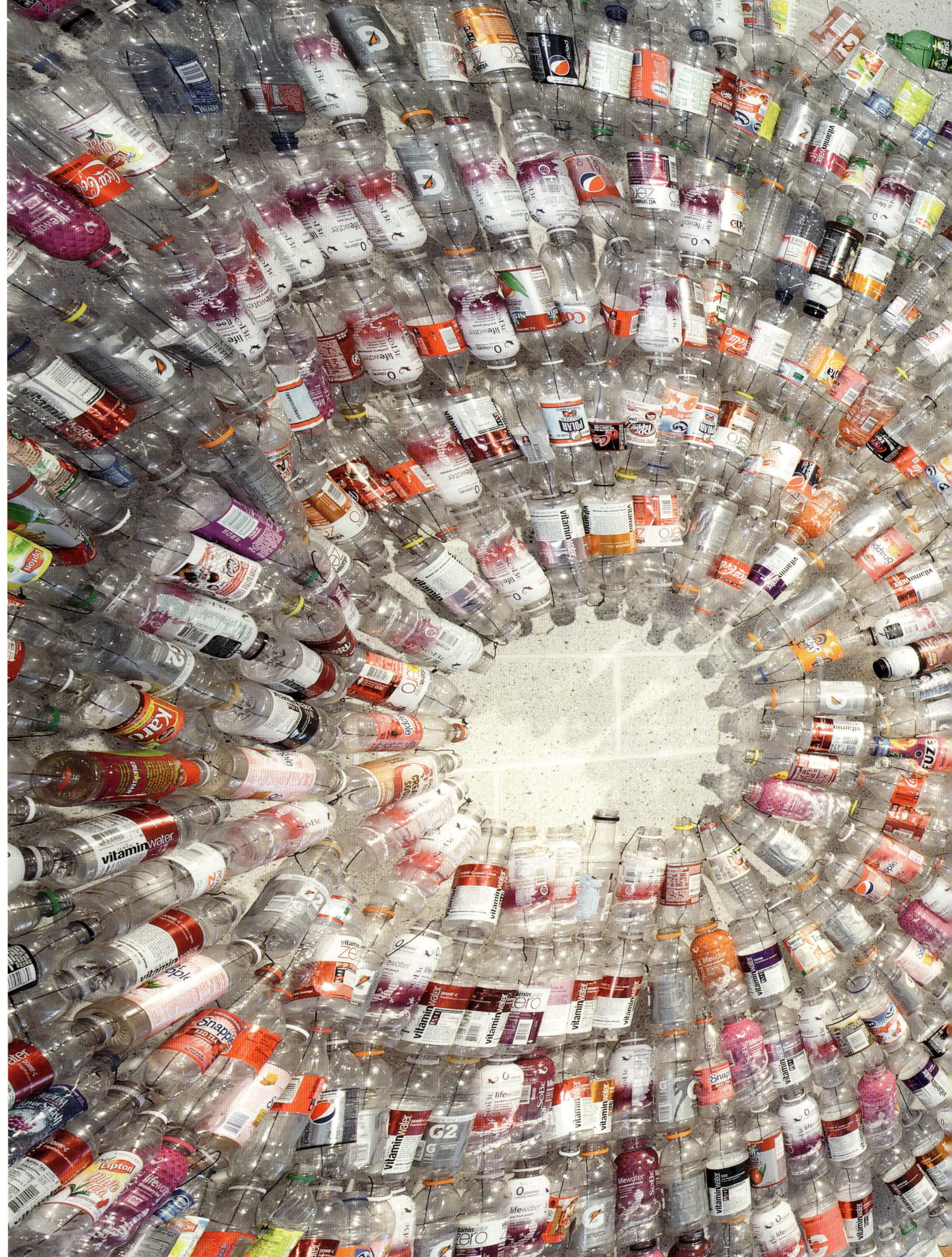

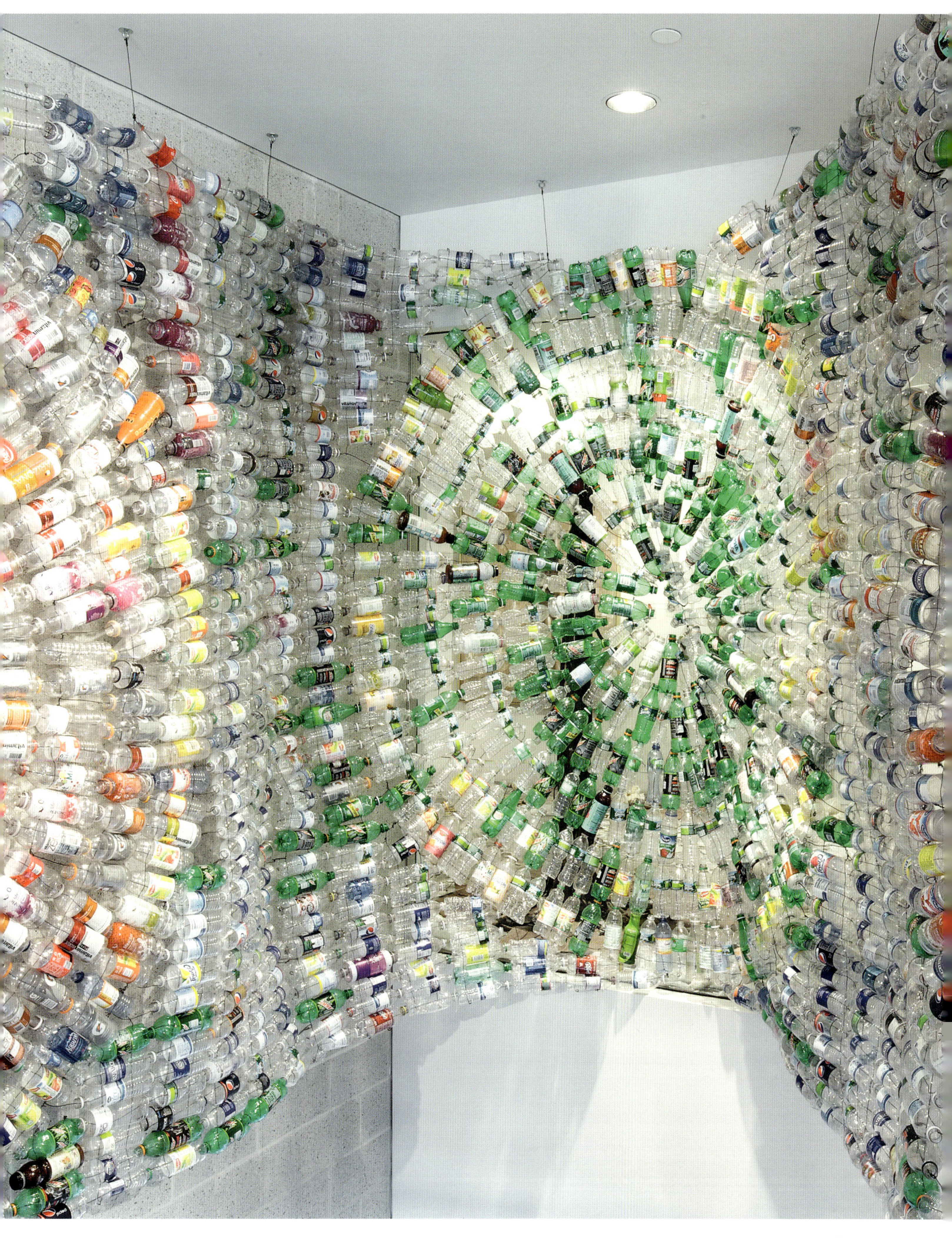

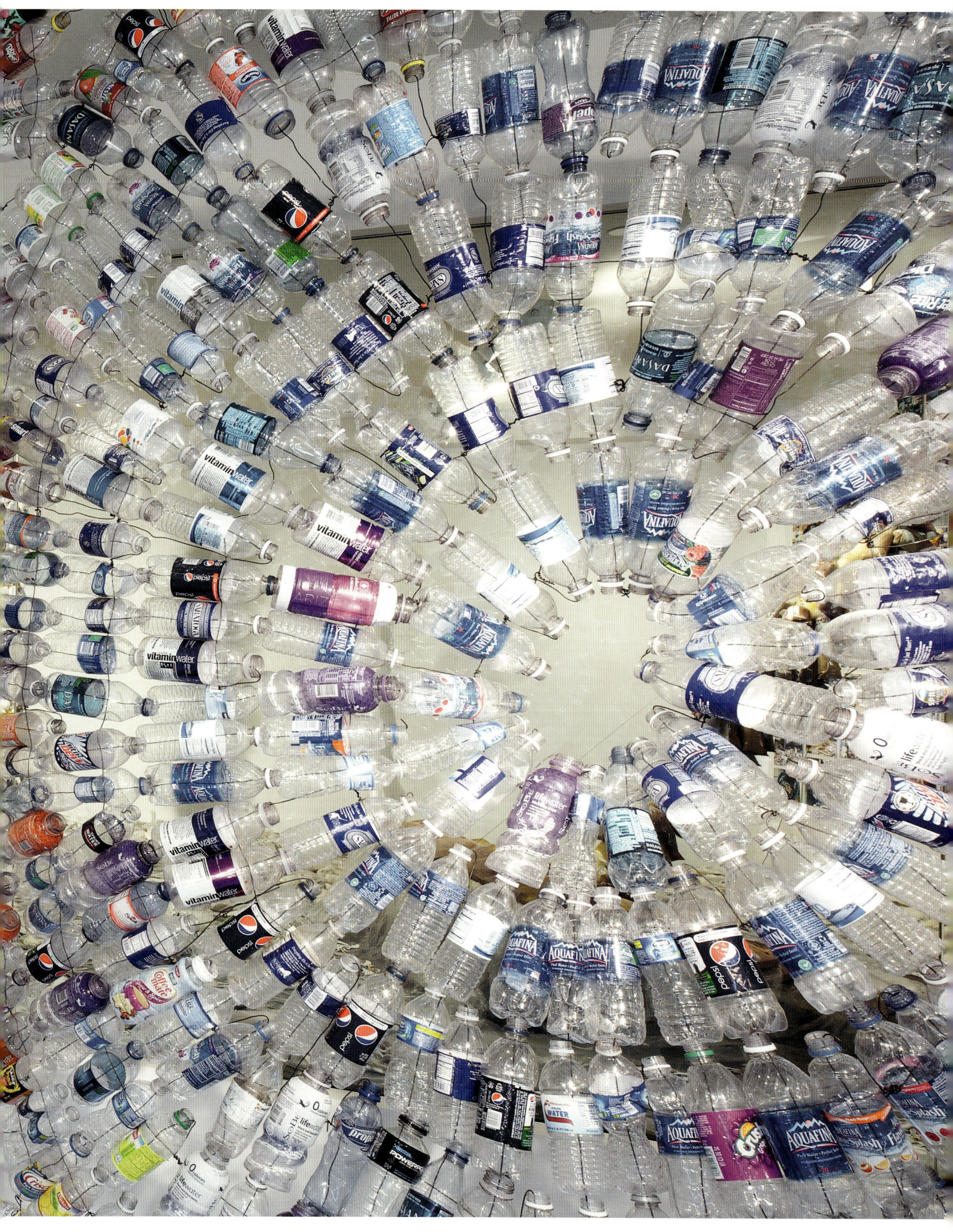

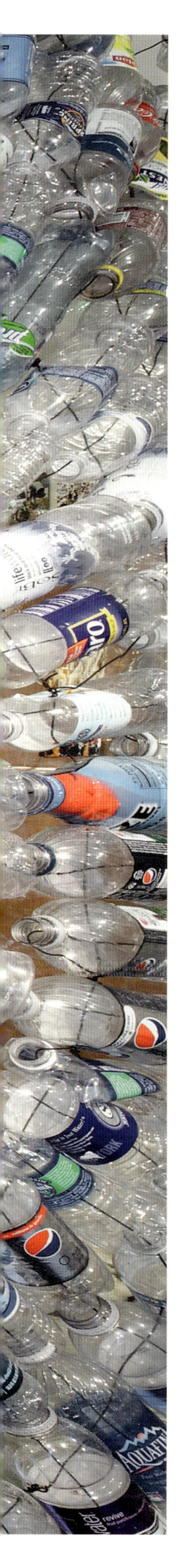

This publication accompanies the exhibition *Environment and Object · Recent African Art* curated by Lisa Aronson and John S. Weber.

The Frances Young Tang
Teaching Museum and Art Gallery
Skidmore College
Saratoga Springs, New York
February 5–July 31, 2011

Anderson Gallery
Virginia Commonwealth
University
Richmond, Virginia
September 9–December 11, 2011

Museum of Art
Middlebury College
Middlebury, Vermont
January 27–April 22, 2012

WEBSITE
http://tang.skidmore.edu/
index.php/calendars/view/298/
tag:1/year:all
The *Environment and Object*
website complements the
gallery exhibition and catalogue,
providing additional historical
information on themes explored
in the artwork, background on
the sixteen artists in the
exhibition, and supplemental
audio and video on the artists and
exhibition.

© 2012 The Frances Young Tang
Teaching Museum and Art
Gallery at Skidmore College, and
Prestel Verlag, Munich · London ·
New York

Editors: Lisa Aronson and
John S. Weber
Design: Barbara Glauber, Azusa
Kobayashi, Gregory Skiano/
Heavy Meta, New York
Printing and binding: Shapco
Printing Inc.

Published in 2012 by The
Frances Young Tang Teaching
Museum and Art Gallery at
Skidmore College, and DelMonico
Books, an imprint of Prestel
Publishing.

The Frances Young Tang
Teaching Museum and
Art Gallery at Skidmore College
Skidmore College
815 North Broadway
Saratoga Springs, NY 12866
T 518 580 8080
F 518 580 5069
www.skidmore.edu/tang

Prestel, a member of
Verlagsgruppe Random House
GmbH

Prestel Verlag
Neumarkter Strasse 28
81673 Munich
Germany
Tel.: 49 (0)89 4136 0
Fax: 49 (0)89 4136 2335
www.prestel.de

Prestel Publishing Ltd.
4 Bloomsbury Place
London WC1A 2QA
United Kingdom
Tel: 44 (0)20 7323 5004
Fax: 44 (0)20 7636 8004

Prestel Publishing
900 Broadway. Suite 603
New York, NY 10003
Tel: 212 995 2720
Fax: 212 995 2733
E-mail: sales@prestel-usa.com
www.prestel.com

All rights reserved. No part of
this book may be reproduced
or transmitted in any form or by
any means, electronic or
mechanical, including photocopy,
recording, or any other
information storage and
retrieval system, or otherwise
without written permission
of the publisher.

LIBRARY OF CONGRESS CATALOGING-IN-PUBLICATION DATA
Environment and object : recent African art / edited by
Lisa Aronson and John S. Weber.
p. cm.
This publication accompanies the exhibition Environment and
Object · Recent African Art curated by Lisa Aronson and John S. Weber.
ISBN 978-3-7913-5209-1 (alk. paper)
1. Art, African—21st century—Exhibitions. 2. Environmental
degradation—Africa—Exhibitions. 3. Landscapes in art.
I. Aronson, Lisa. II. Weber, John S.
N7380.5.E58 2012
709.6—dc23
2011041187

British Library Cataloguing-in-
Publication Data: a catalogue
record for this book is available
from the British Library; Deutsche
Nationalbibliothek holds a record
of this publication in the Deutsche
Nationalbibliografie; detailed
bibliographical data can be found
under http://dnb.ddb.de.

The paper in this book meets
the guidelines for permanence
and durability of the Committee
on Production Guidelines for
Book Longevity of the Council of
Library Resources.

FRONT COVER:
Romuald Hazoumé
Claudia Maigre, 2005
Found objects, plastic, and
synthetic and cotton thread
11½ × 10½ × 8 inches

BACK COVER:
Barthélémy Toguo
*Stupid African President 3
(Afrika Oil)*, 2006
Digital inkjet print
42⅞ × 30⅞ inches

PAGES 4–5, 6:
Bright Ugochukwu Eke
Ripples and Storm 1, 2011
(details)
Recycled bottles and wire
Installation dimensions
variable

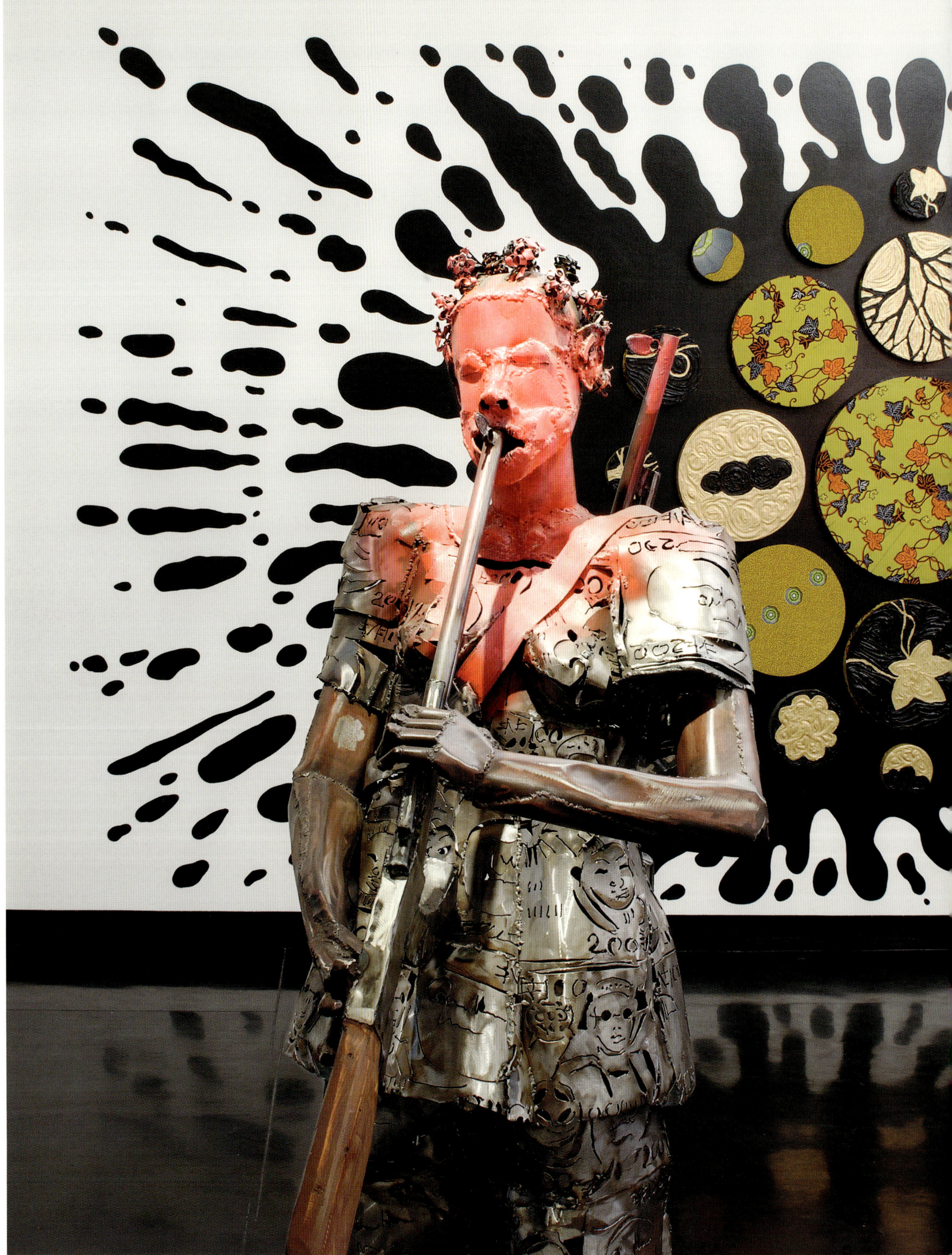

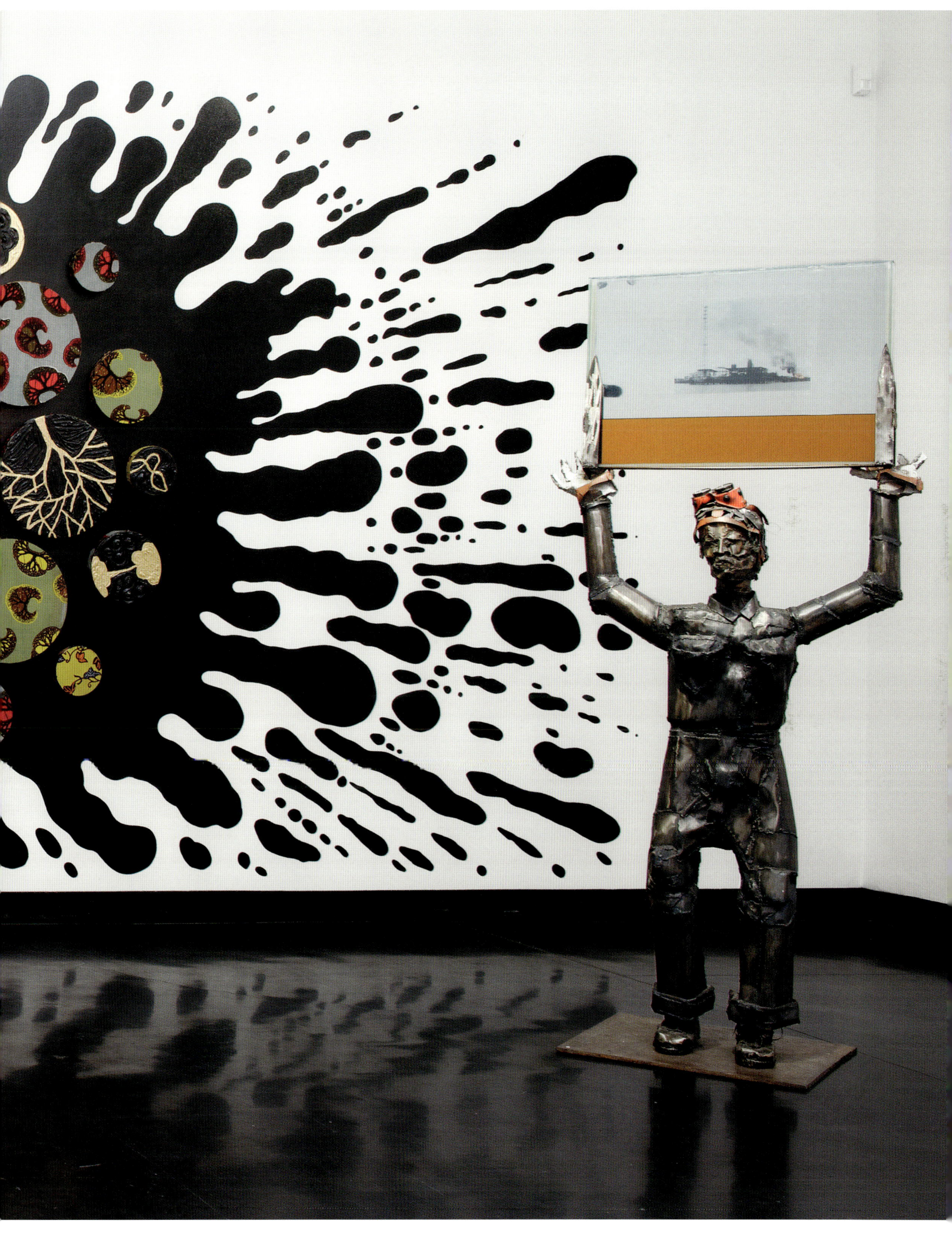

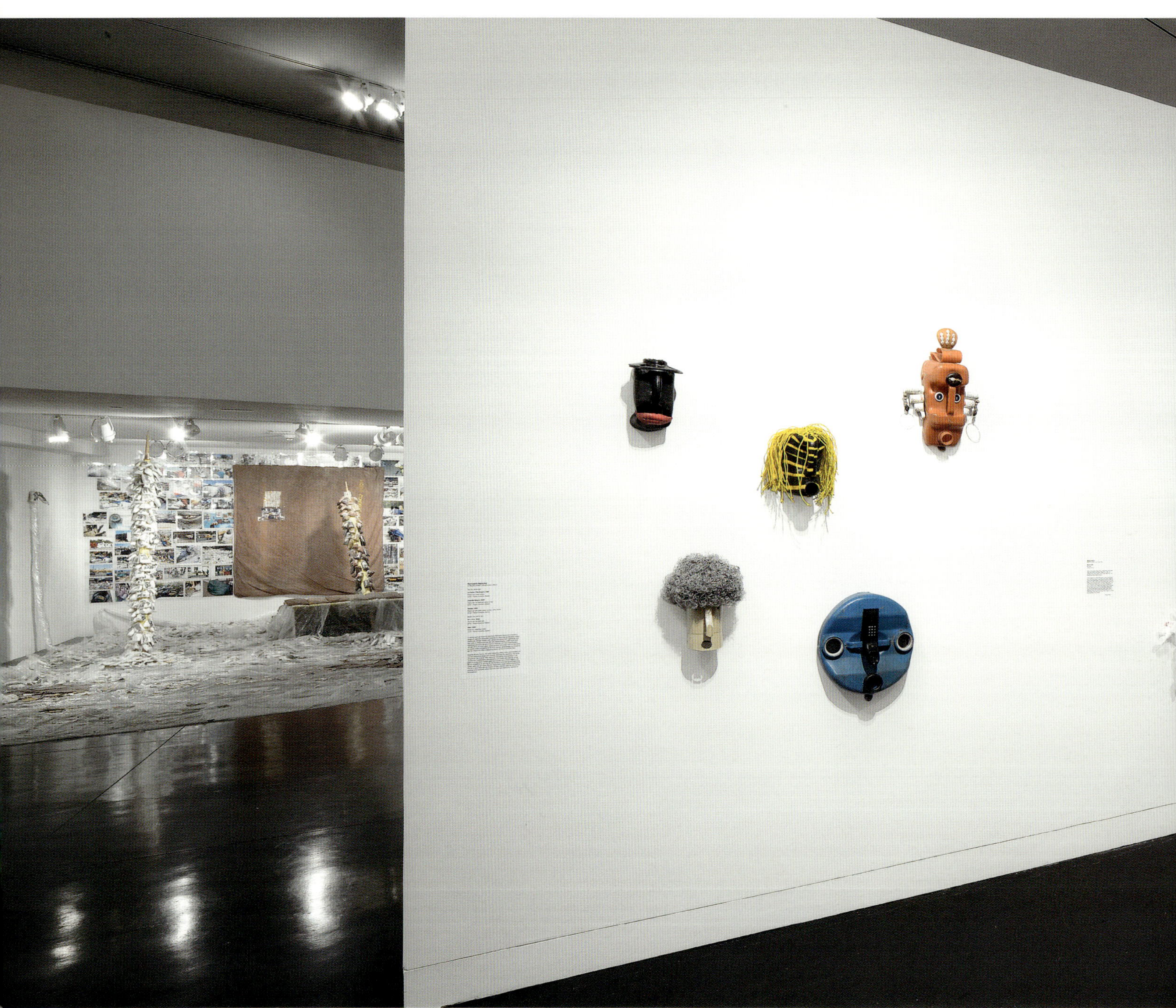

VIYÉ DIBA,
ROMUALD HAZOUMÉ,
AND NNENNA OKORE
Installation view,
*Environment and Object ·
Recent African Art*,
Tang Museum, 2011

PREVIOUS SPREAD
SOKARI DOUGLAS CAMP
AND YINKA SHONIBARE
Installation view,
*Environment and Object ·
Recent African Art*,
Tang Museum, 2011

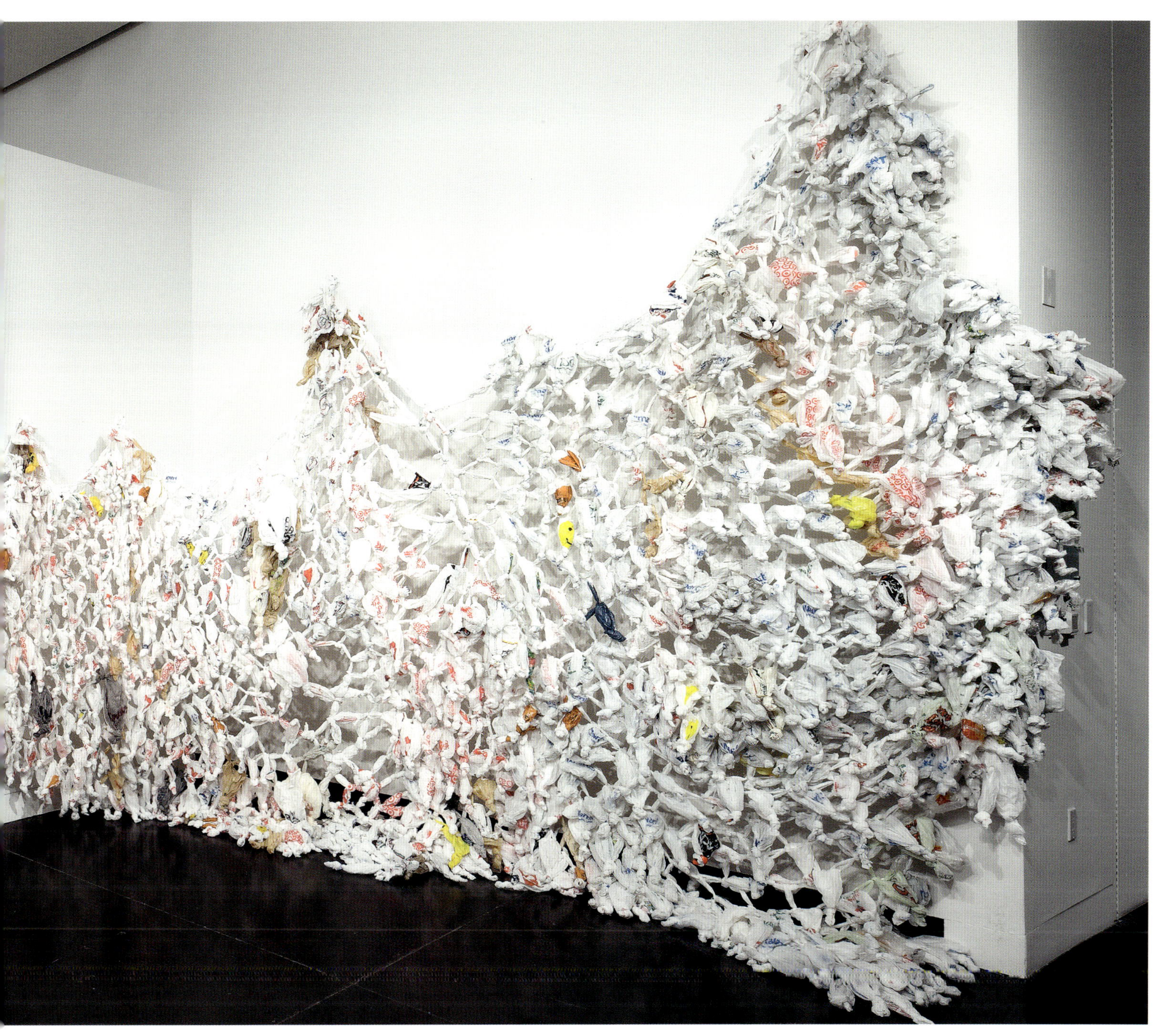

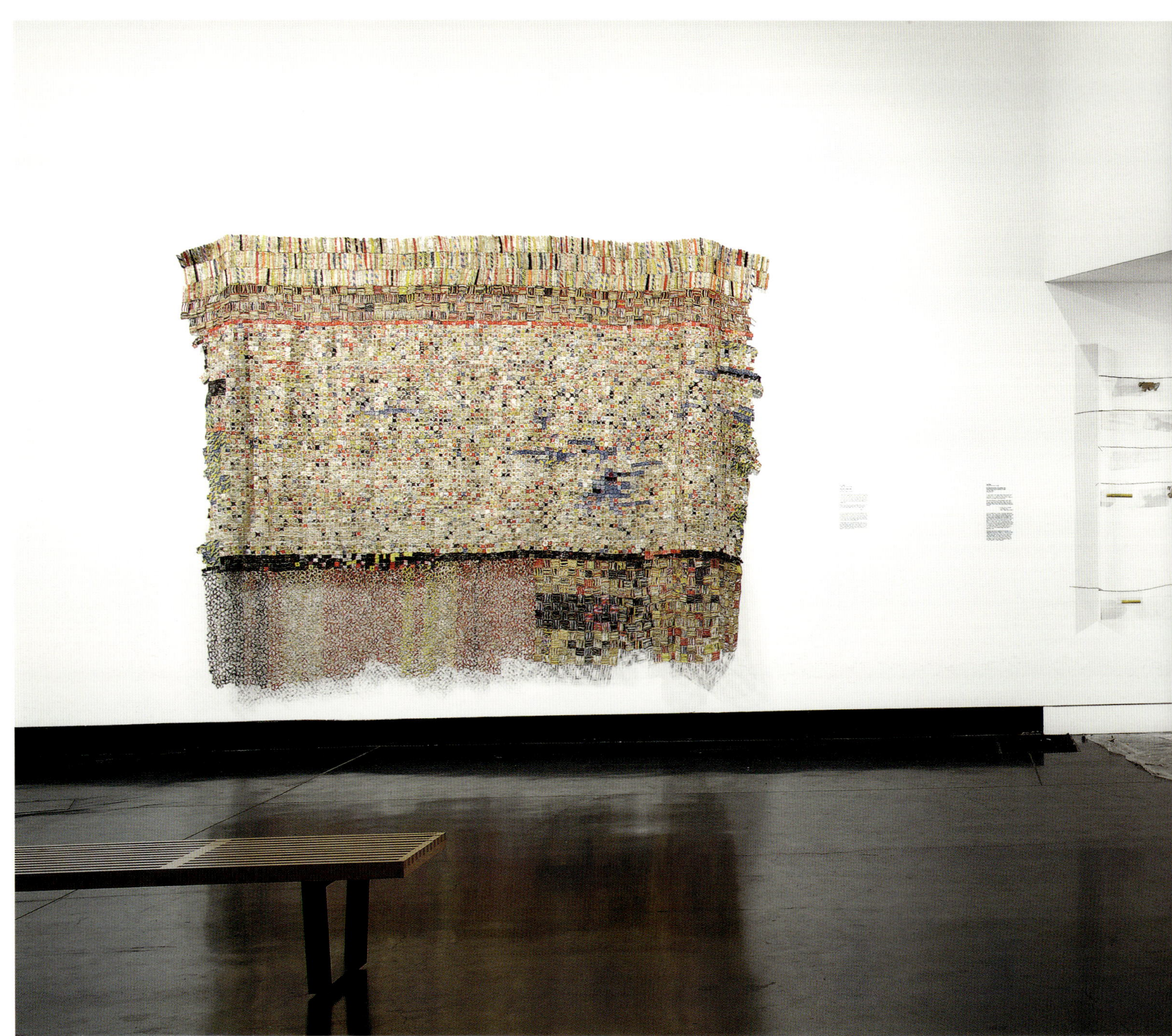

EL ANATSUI, VIYÉ DIBA,
AND BARTHÉLÉMY
TOGUO
Installation view,
*Environment and Object ·
Recent African Art*,
Tang Museum, 2011

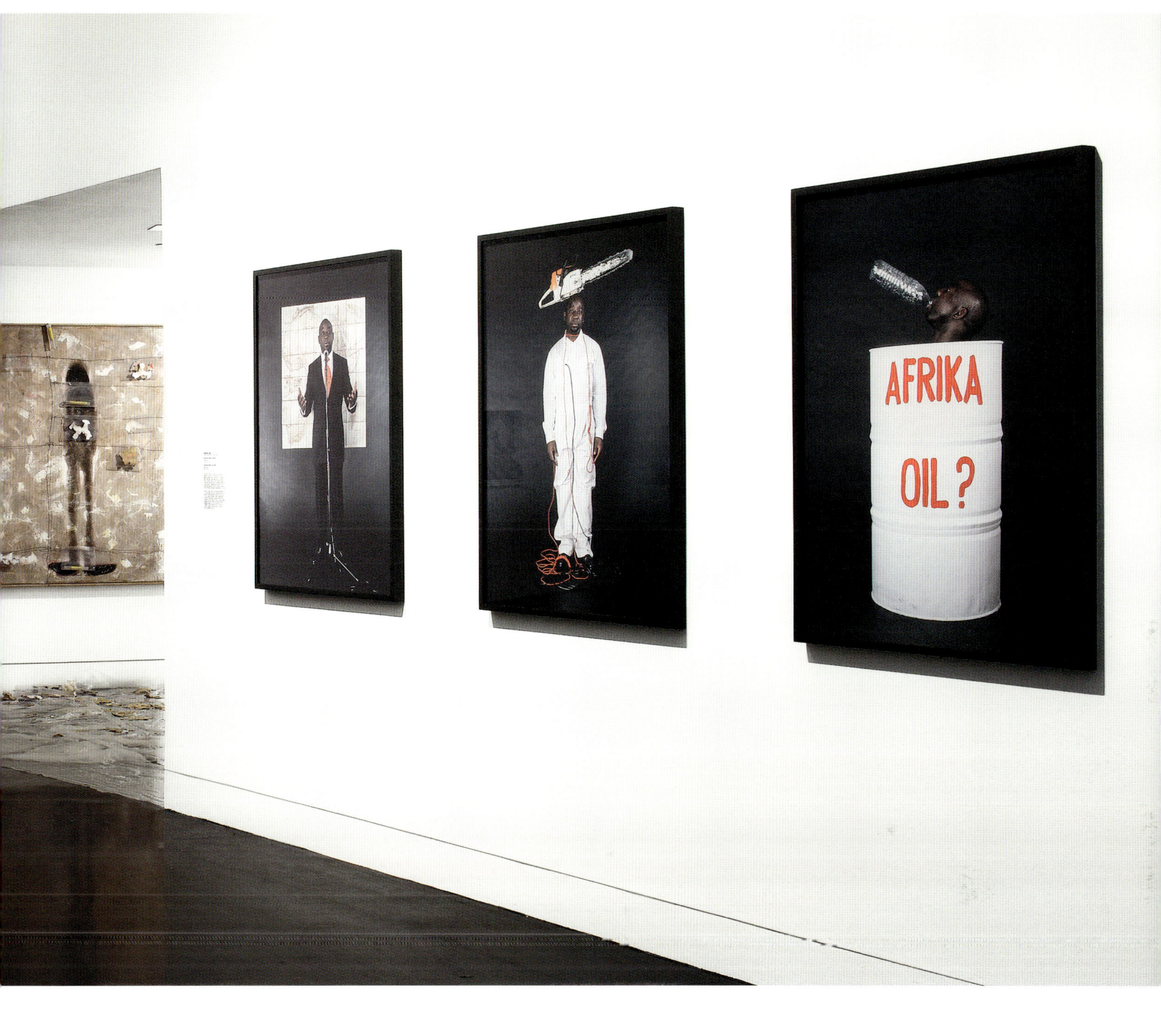

AFRIKA
OIL ?

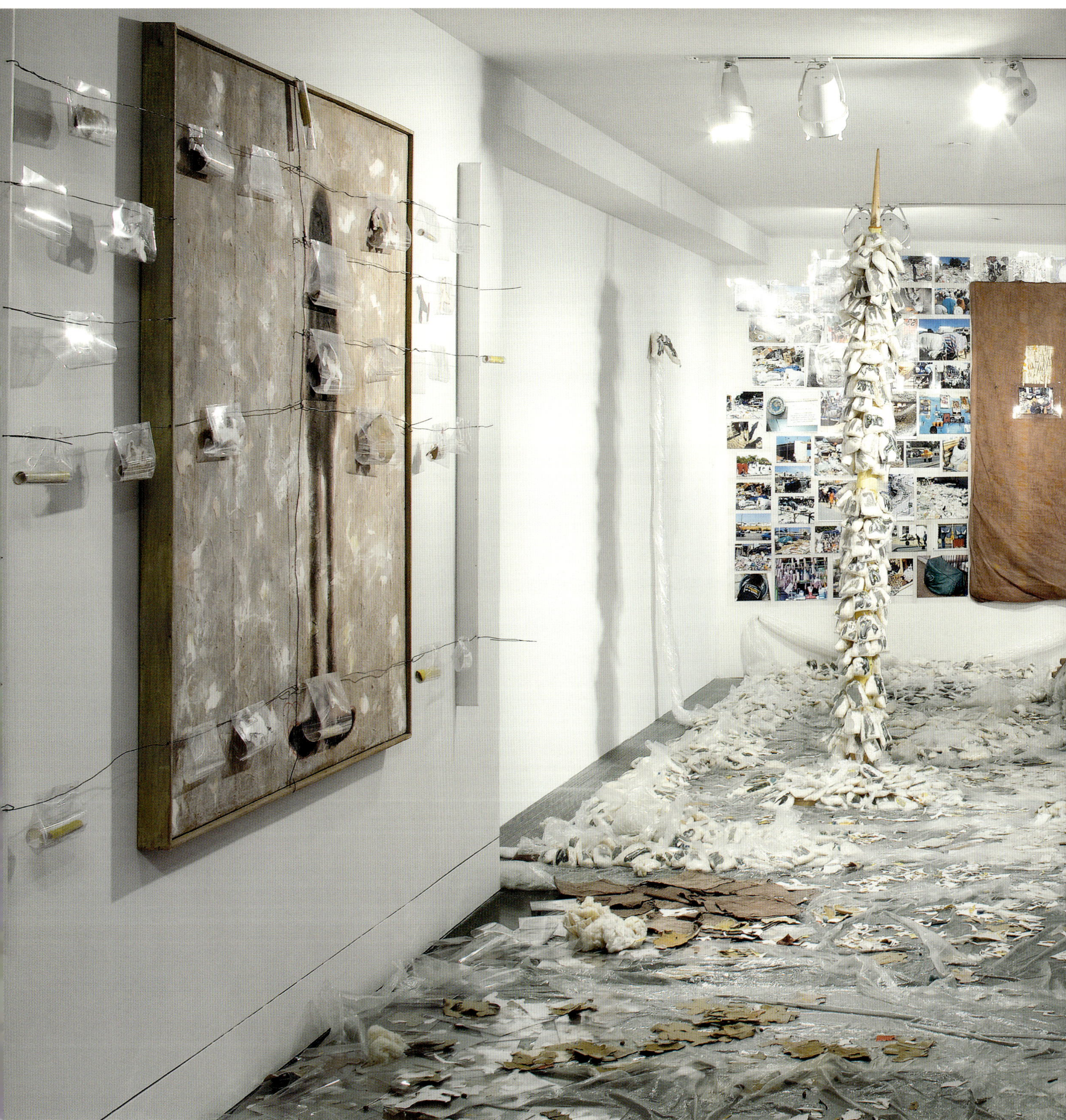

VIYÉ DIBA
Nous sommes nombreux,
et nos problèmes avec…
[We are numerous,
and our problems with
that…], 2008
Mixed media
Installation dimensions
variable

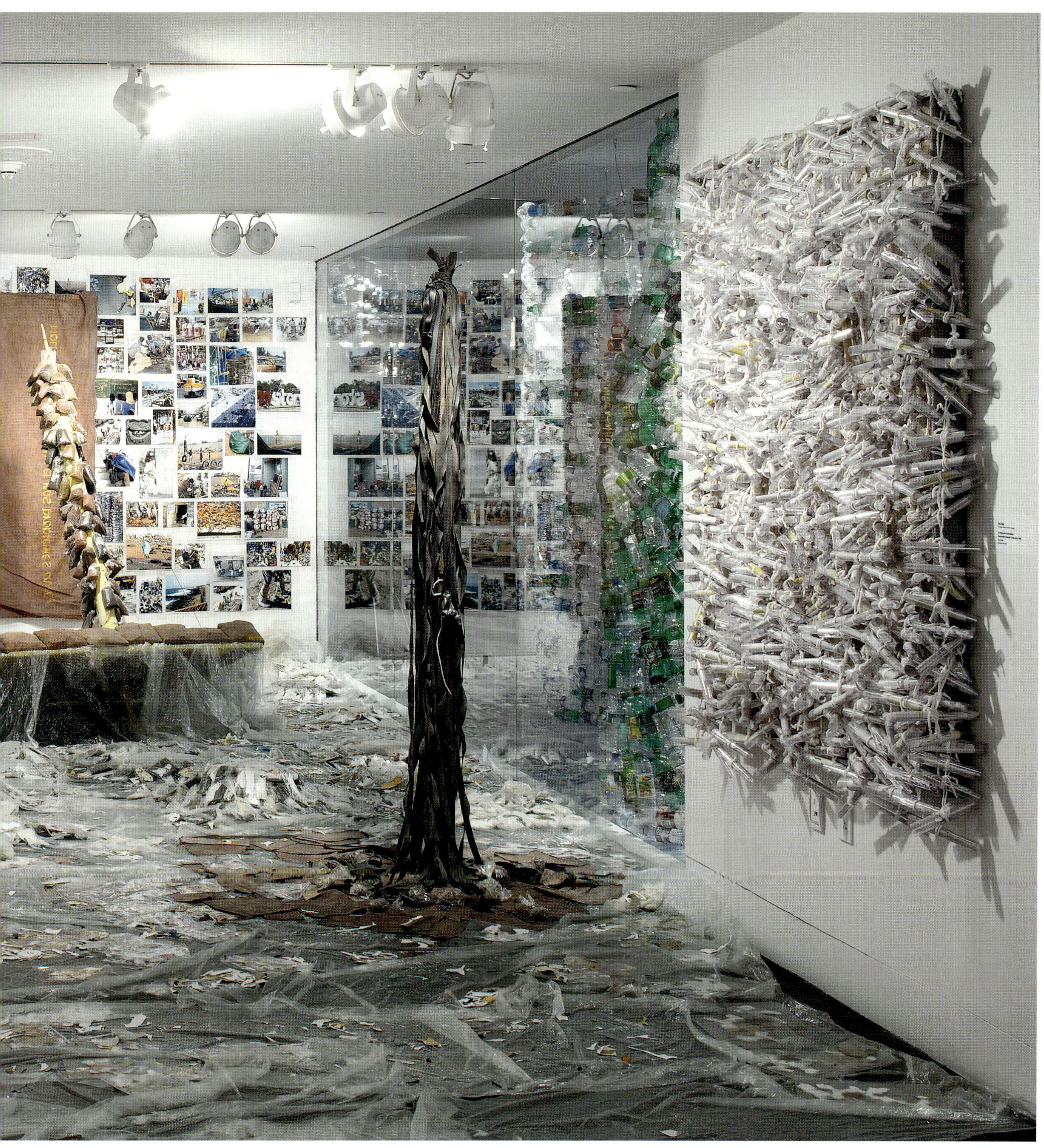

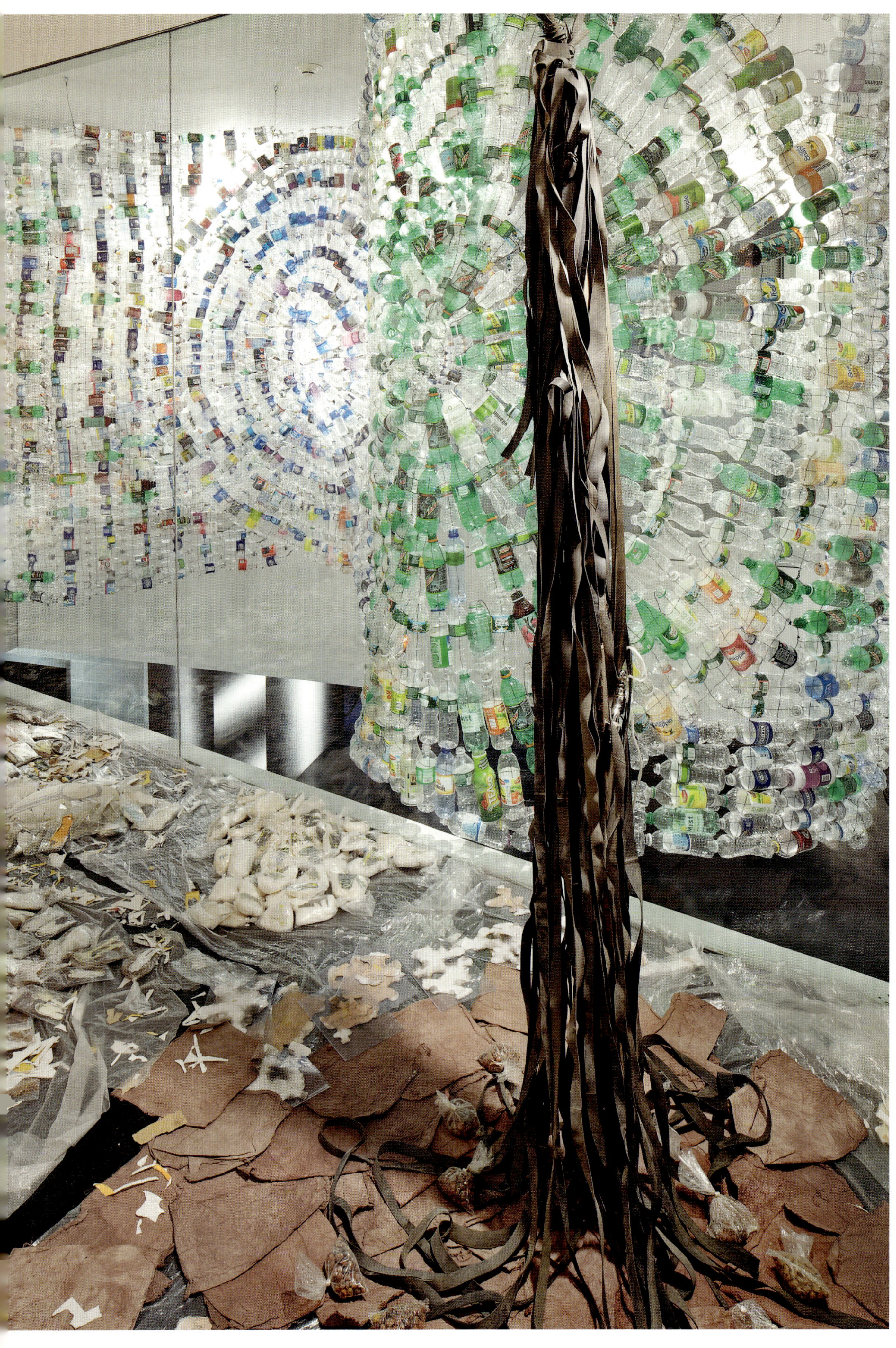

Introduction

More than three years in the making, *Environment and Object · Recent African Art* joins a remarkable recent upwelling of interest in contemporary art from Africa and African artists. Over the past decade, African artists have participated regularly in major international surveys and biennials, and it has become more common, though still relatively novel, to see in such exhibitions work by artists from Nigeria, South Africa, Senegal, Egypt, Cameroon, or one of the other forty-nine nations on the African continent. Whether reflecting or driving this trend, exhibitions and sales of work by contemporary African artists have also increased dramatically in the galleries of New York, London, Paris, and other art market centers in recent years, and art museums are eagerly acquiring contemporary African works.

Environment and Object · Recent African Art participates in this trend but also represents a comparatively new development, namely, the examination of distinct themes and issues in the field. Our title signals the conceptual organizing principle of the exhibition, which looks at how African artists tackle environmental topics in their work and respond to their natural, urban, and economic environments in their selection of media and working materials. At one pole, such artists as Yinka Shonibare, in *Black Gold I* and *Black Gold II*, or Sokari Douglas Camp, in *Teasing Suicide*, draw attention to the oil industry's horrific environmental and social damage to the Niger Delta, employing alternately more abstract (Shonibare) and more literal (Douglas Camp) modes of address. At *Environment and Object*'s other conceptual pole, the found objects employed by

    Lisa Aronson and John S. Weber

El Anatsui in *Some Still Come Back* and Nnenna Okore in *Mbembe*
echo the material realities of their environments, implicitly
commenting on the varied historical economies of alcohol (Anatsui)
and plastic (Okore).

In developing the central idea and selecting works for *Environment
and Object*, we sought to follow the artists' lead and not simply
impose a theme on an existing and unusually rich field of aesthetic
activity. Indeed, the exhibition was originally conceived as a survey of
recent African art that would use metaphors and forms of the body
as its guiding theme. A trip to the Dak'Art 2008 biennial and a
survey of recent exhibitions and publications made it clear that "the
body," however legitimate as a topic, represented a well-trodden
curatorial and critical path. In contrast, the complex yet unmistakable
presence of the environment in recent work from Africa warranted
more thought, investigation, and exploration. Engaging artists in a
wide range of ways, environmental issues and situations have
clearly motivated some of the most exciting new work and offered
a promising theme around which to mold a compelling exhibition.

We found strikingly original artworks that interrogated a variety
of environmental issues, ranging from the more strictly ecological (oil
problems, acid rain, access to clean drinking water) to the urban
and social (urban overcrowding, lack of civic infrastructure).
Speaking with artists, curators, dealers, scholars, and now the public
for *Environment and Object*, we have stressed the wide scope of
concerns and engagements evident in the artworks on view and the
expansive notion of "environment" in play here. Perhaps most
significantly, by invoking environmental questions as a key
organizing principle, we have highlighted artists' evident and urgent

concerns regarding specific situations and conditions of the external world, the natural and human-made contexts within which these artists live, and the problems they see all around them.

The term *object* in our title refers most prominently to the physical objects from which a number of the works in the exhibition are fashioned. These "found objects" had a life and a history previous to their recuperation as raw material for art. As such, they speak to and carry those histories with them into the artworks they become, bringing with them traces of their earlier situations and conditions—of their environments. Rejecting traditional and essentially anonymous modern art materials, such as oil paint and canvas, many of the artists in this exhibition employ found objects in art making as a way to locate their creative work within specific environments and historical situations. Hence, the use of found objects represents a political as well as an aesthetic choice.

The use of found materials and the relationship of art to its environment both possess a long history in African art, and as we conceived of the exhibition, we originally envisioned showing traditional work—*Ijo owu* water-spirit masquerades, Yoruba Osanyin staffs associated with herbal healing, or Bamana Chi-Wara masks celebrating productive agricultural work—alongside the contemporary. Conversations with other scholars and curators of recent African art and feedback from artists themselves convinced us that the exhibition would be stronger if confined to contemporary works alone. The inclusion of traditional work, we became convinced, risked situating contemporary artists first and foremost in relation to assumed and ascribed roles as inheritors of tribal and ethnographic traditions. Given the varied, subtle, and at times ambivalent

 Lisa Aronson and John S. Weber

relationships that contemporary African artists have with their competing identities as "African" and "contemporary" artists, anchoring their work in the gallery in relation to village- and ethnic-based art traditions did not seem wise. The fact that *Environment and Object* focuses dramatically on contemporary issues also argued strongly for the exclusion of traditional objects from the galleries. That said, we gladly and unapologetically embrace historical and anthropological approaches in both the catalogue and website for *Environment and Object*.

Chika Okeke-Agulu notes in his insightful interview for this catalogue that tension between anthropological and art historical treatments of African art has been an issue in the field that will likely continue. Contemporary African artists play an ever-larger role in the global art world, attracting new generations of critics, scholars, collectors, and dealers. Attracted to this work on the basis of its contemporary feel, topical relevance, and market visibility, these newcomers may have little or no background in traditional African art history. Yet the increased interest in art produced by contemporary African artists will, we believe, only increase awareness and familiarity with African histories and traditions, which can only benefit the larger field. Undoubtedly, the art market currently supports—and in some cases is infatuated with— recent African art, which now joins a long list of contemporary, international market fads: Italian and German art in the early 1980s, identity-based political art in the early 1990s, Japanese and Chinese art since the late 1990s, and so on. A pessimistic view of these developments would condemn them as a new form of colonialism, with rich nations seeking new cultural resources to exploit. A more optimistic view would consider that once the hype and market fads fade, what remains, as with the work of contemporary artists from other regions, will surely be an expanded field of critical inquiry and a greater openness to artists seeking to show their work internationally. Whichever view one takes, it is too early to predict the ending of this new African story. Even less clear is what effect the current art market enthusiasm for these works in Europe and the United States will have on opportunities in African nations themselves for artists to show, sell, and live from their work, an issue beyond the scope of this exhibition or introduction.

In shaping this exhibition, we envisioned the show and this catalogue as distinct but interdependent entities and the website as a complement to both. The catalogue and website contextualize the work itself, and the catalogue also visually documents the exhibition's form and content. In our two essays, we consider how artists both comment on and draw from the African landscape. Furthermore, essays by environmental scientist Karen Kellogg, political scientist Christopher Whann, and sociocultural anthropologist Mark Auslander introduce nonspecialist readers to historical conditions that have helped shape the artists' work, thereby expanding its impact and accessibility. Chika Okeke-Agulu's interview offers a perceptive and highly informative overview of the broader field of modern and contemporary African art.

By spotlighting work related to environmental issues, we also hope to turn a conceptual page in the field of contemporary African art. To date, many exhibitions of recent art from Africa have offered broad surveys, essentially samplers of work from the continent and its artists. Others have hewn more closely to topics typically related to traditional African art, such as the body. As such, they have

          Lisa Aronson and John S. Weber

primarily, if at times unintentionally, foregrounded the "Africanness" of the work and artists on view. We conceived of *Environment and Object* in a different way. Following the lead of the artists and works themselves, we see the exhibition as an investigation of environmental situations relevant to the world as a whole, regardless of whether the site of artistic interrogation is in Africa or elsewhere. Within this context, we have reframed the view of such artists as Yinka Shonibare and El Anatsui, focusing less on their African identity than on their works' relationship to transnational questions and issues. To be sure, *Environment and Object* is about Africa, yet it emphasizes points of connection and shared challenges that embed Africa in the global environment—ecologically, economically, and socially.

The art and artists on view in *Environment and Object · Recent African Art* ask us to reexamine, in the words of Bright Ugochukwu Eke, "the Human-Earth connection" in Africa and wherever else we may live. They ask us to rethink our relationship to the objects and substances we consume daily, where they come from, and who harvests the profits they earn. They ask us to question the nature of our cities and how we live together in them. Finally, they ask us to become aware of the histories that link Africa to the rest of the world and how those histories continue to shape African environments today.

Lisa Aronson, *Associate Professor of Art History*
John S. Weber, *Dayton Director, The Frances Young Tang Teaching Museum and Art Gallery*

Skidmore College, Saratoga Springs, New York

Lisa Aronson

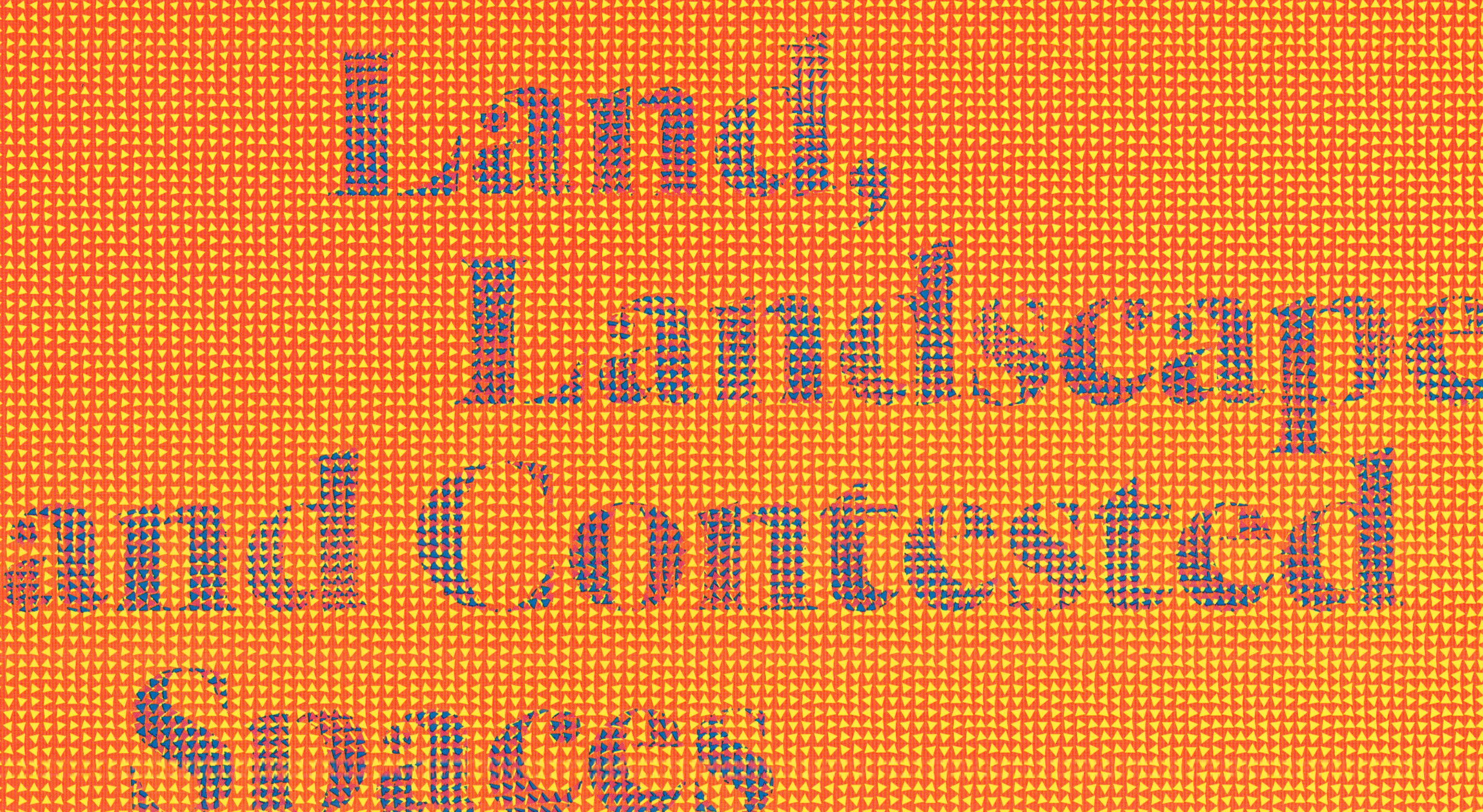

Oil spillage for me is both a literal and metaphoric way of interrogating the myriad and complex way in which humanity as a whole has systematically destroyed planet Earth by its destructive activities.[1]

rtist **Jerry Buhari**'s words about his painting *In the Pipeline* capture the spirit behind many of the works in the Tang exhibition *Environment and Object · Recent African Art.* Decisively rejecting Western, romanticized perceptions of the African landscape as an isolated domain of lush, green jungles and vast savannahs filled with wild animals, these artists instead interrogate contemporary African landscapes, both urban and natural, as contested and compromised spaces of economic, social, and political dimension. The art in this exhibition touches on everything from oil, water, and minerals to deforestation, soil depletion, poor trash management, and malarial mosquito infestation. Moreover, it addresses human elements and the built environment as well as the natural inland scenery, with its flora, fauna, and landscape contours.

The sixteen artists in this show either come from Africa or claim African heritage, and while many live and work abroad, their works arguably feel African, whether through conscious references to "tradition," the ironic humor of their messages, or, most important, their engagement with the many African landscapes as spaces to be reckoned with. The artists create in a variety of artistic styles and modes of expression, including photography, sculpture, painting, and installation, offering works from the most abstract or conceptual to the most realistic. The exhibition's great variety and diversity remind us that there is no one narrative of contemporary African art.

"Landscape" Art History

Artists training in some areas of post-liberation Africa emphasized the study and interrogation of landscape imagery as a primary mode of expression. At Amadu Bello University in northern Nigeria in the 1970s, where Jerry Buhari was then studying art, students took a series of landscape-related courses, such as landscape drawing and landscape painting.[2] For Buhari, these courses instilled "a sense of deep observation of his environment" and an "intimate encounter and dialogue with the landscape"[3] that continue to inform his work.

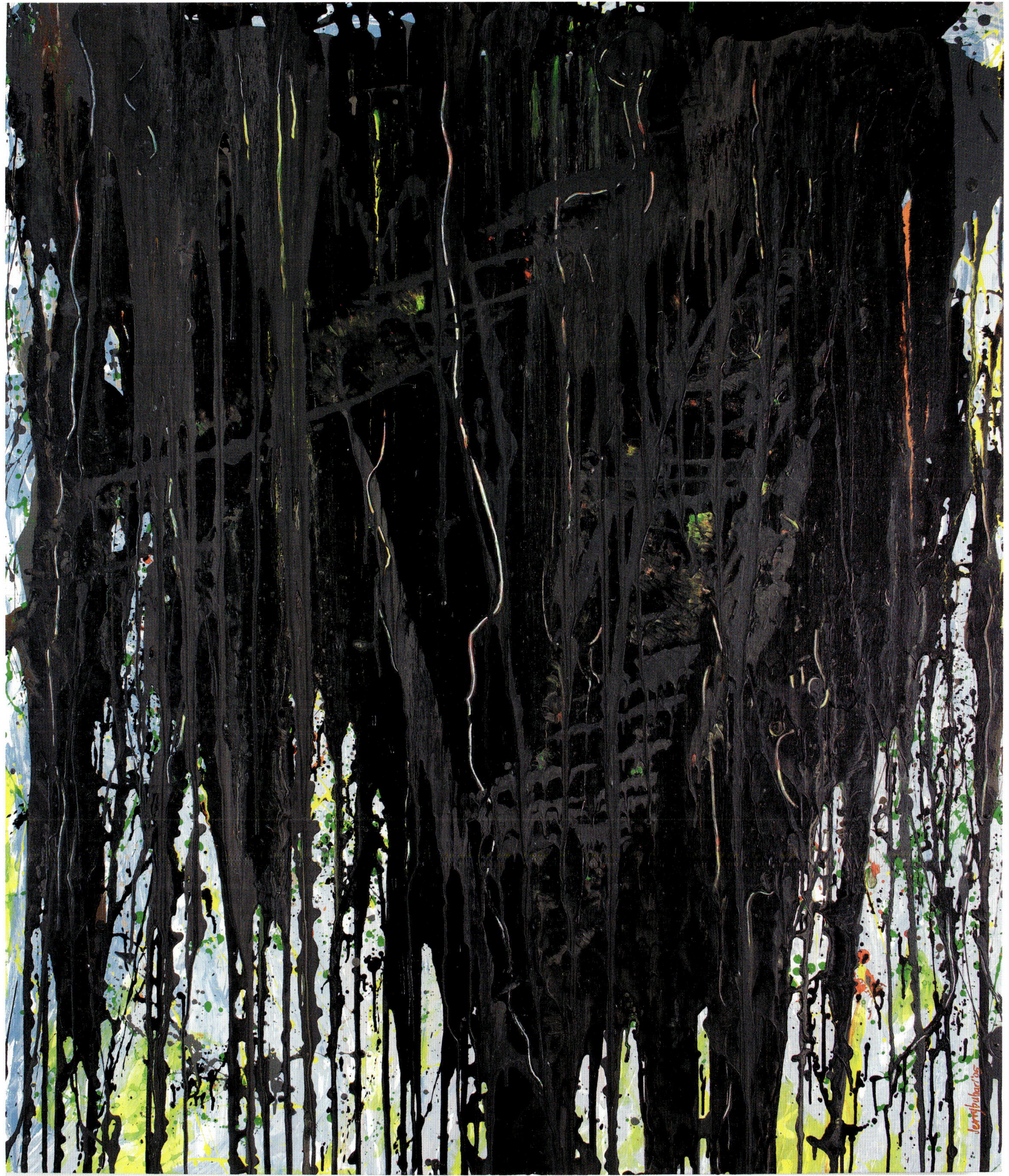

     Lisa Aronson

JERRY BUHARI
Decay and Regeneration III,
2009
Termite-eaten wood, scrap
rubber, and acrylic
8⅞ × 24⅞ × 8½ inches

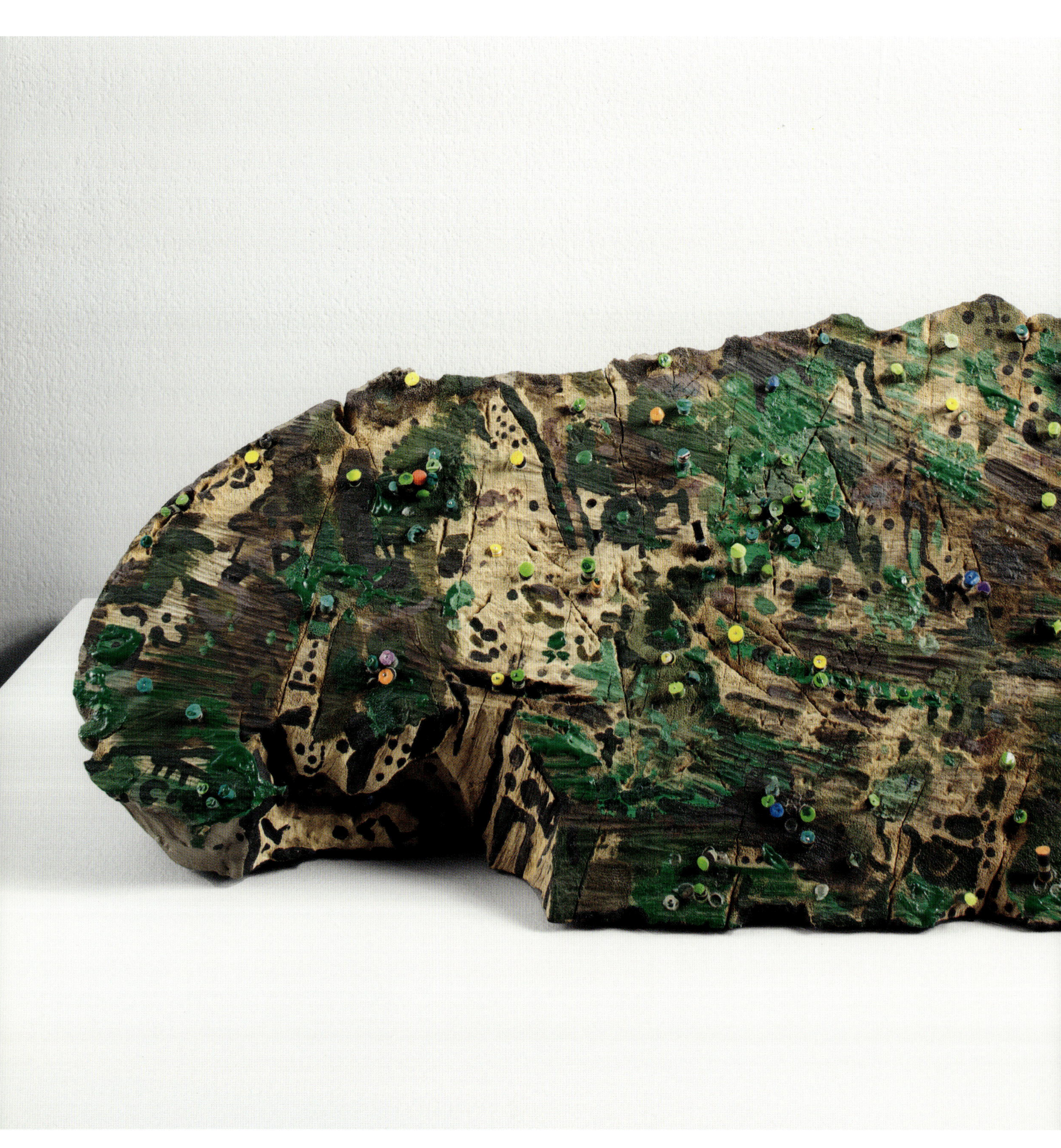

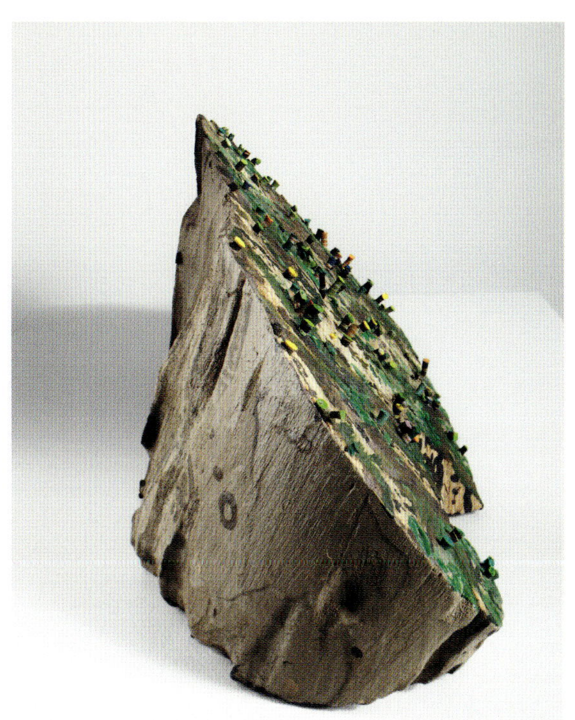

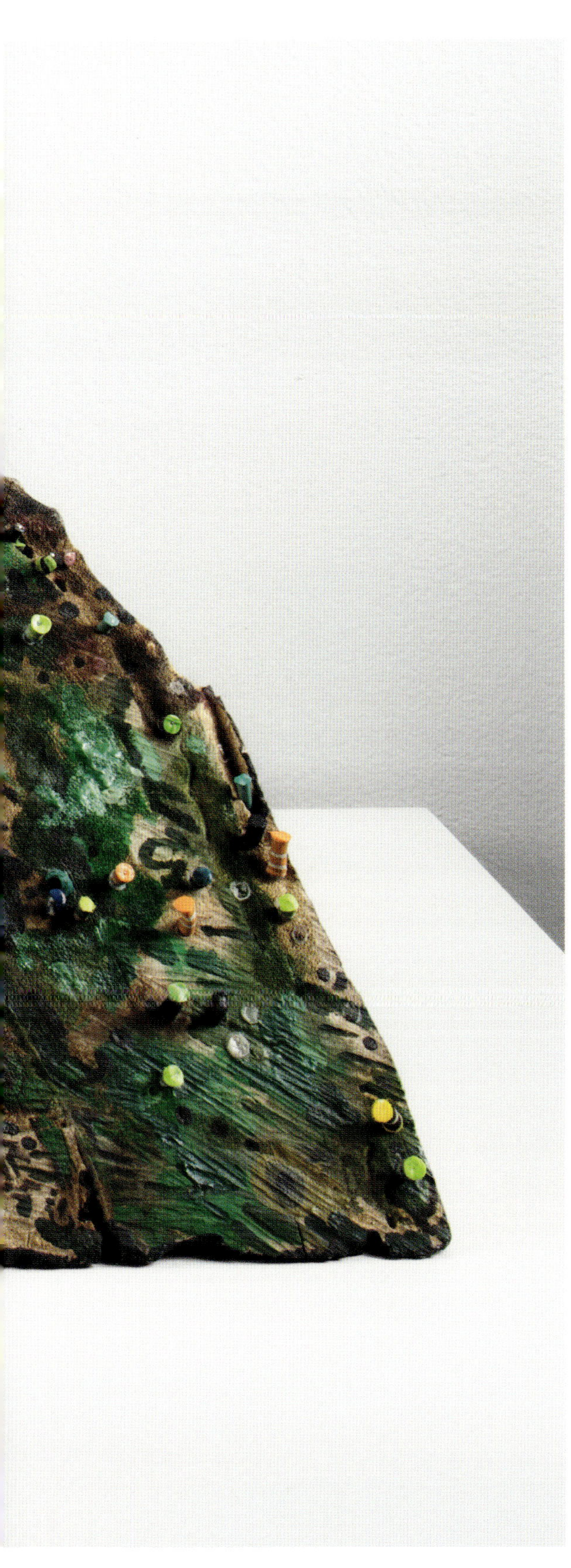

 Lisa Aronson

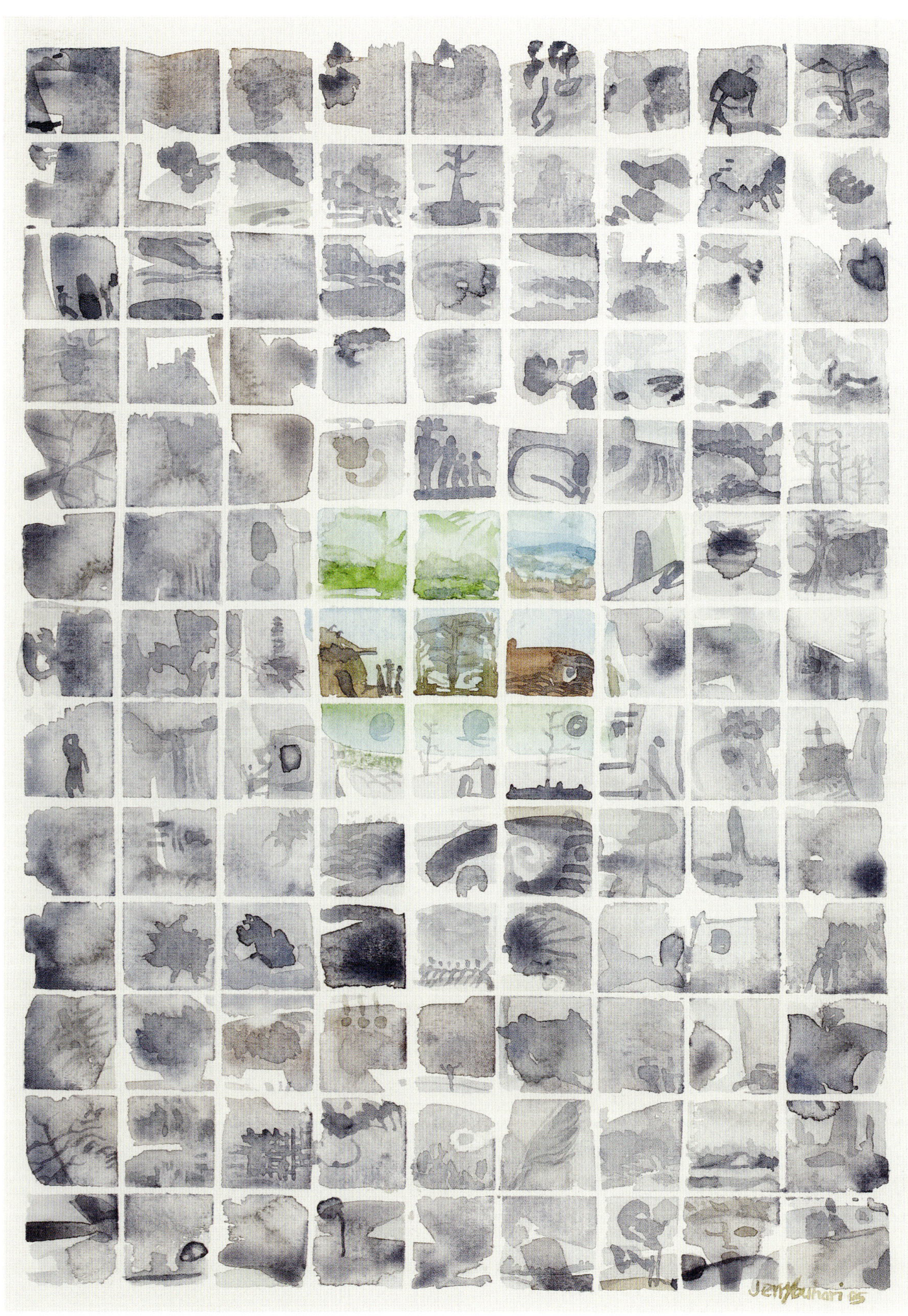

  Land, Landscape, and Contested Spaces

Some of Buhari's works present romanticized views of nature, while others, including those in this show, address landscape and its destruction in a more critical way.[4] His painting *In the Pipeline* (2005) speaks to the oil conflicts in Nigeria's Niger Delta region, now one of the largest sources of crude oil on the entire African continent and the fourth largest in the world. Oil spills happen on a daily basis, leaving pollution and human destruction in their path. Buhari spills black acrylic, a petroleum-based paint, over pastel hues symbolic of nature, drawing our attention to the spills and other consequences of the oil conflict that plague this region.[5]

Buhari's art also concerns the fate of his more immediate environment, northern Nigeria, including its ancient and historic cities. His miniature watercolor, *Images of Zaria City* (2005)—part of a larger series titled *Exploring Images of a City*—comments on the ill effects of slash-and-burn agriculture. In *Images of Zaria City*, he painted 117 miniature landscape views of the Zaria region, only nine of them, the most centrally positioned, in color, and the remaining 108 mostly in black and white. In this way, Buhari implies that few natural areas remain unadulterated and colorful and that the tainted, now dominant landscape rendered in black and white threatens to encroach upon them.

The landscape focus of South African artists **Georgia Papageorge** and **Garth Meyer** harkens back to a long history in their country, even as their work approaches the land with radically different premises and intentions than in earlier landscape art in South Africa. Between 1900 and 1930, British settlers were painting landscape imagery to assert their own identities in their newly claimed territories. Jeremy Foster convincingly argues in his book *Washed With Sun* that these artists' idyllic, picturesque landscapes consistently appeared void of indigenous people, or any human presence for that matter, to clear the way for white ownership of territories that looked wild, while simultaneously suppressing any hint of racial confrontation.[6]

Garth Meyer's landscape photographs turn our attention away from the apartheid-related issues commonly associated with South African politics and toward questions of ecology and environment. Using an 11-by-14-inch format camera, he creates large-scale, black-and-white views of the African forest that are both stunning and haunting. Some have argued that his works formally resemble aspects of Abstract Expressionist

    Lisa Aronson

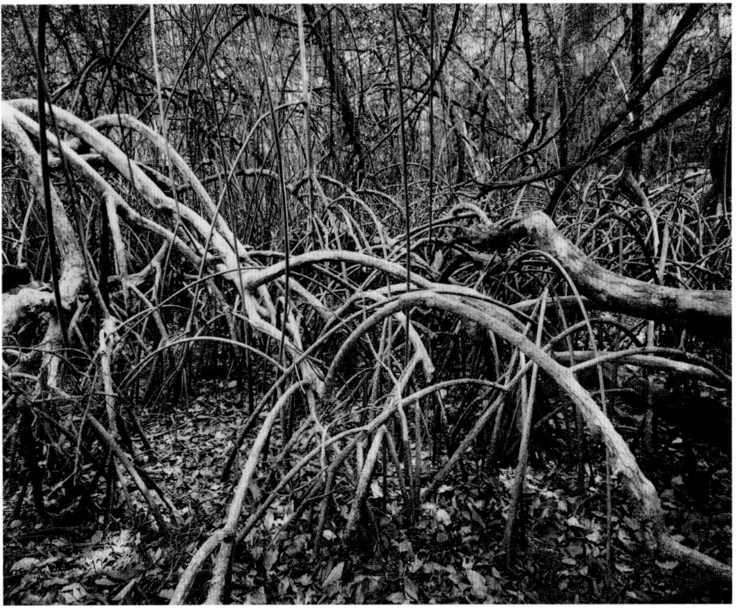

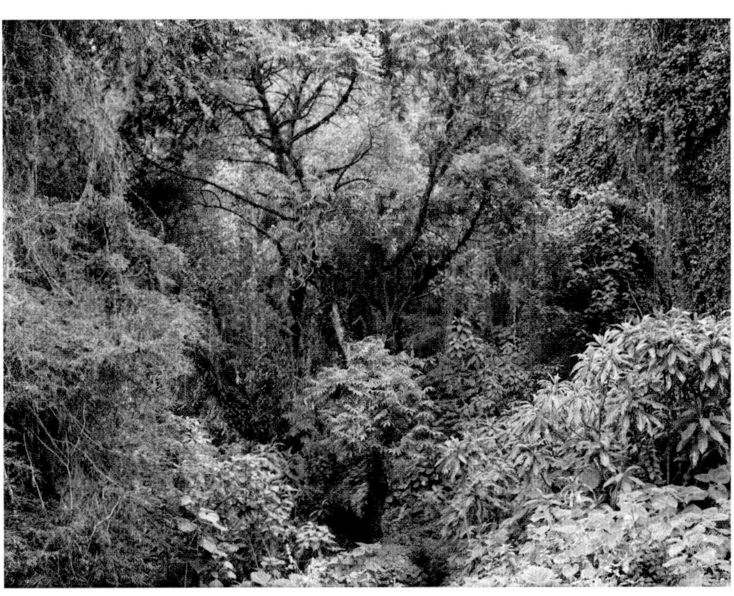

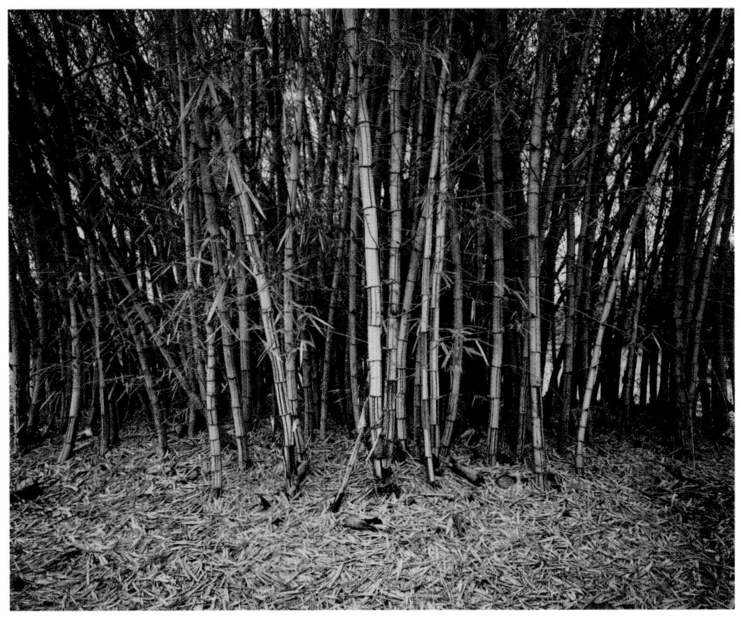

GARTH MEYER

Untitled (Figure #4), 2009
Pangora, Gabon
Silver halide hand print
10¼ × 13⅛ inches

Untitled (Figure #1), 2004
Wild Coast, South Africa
Silver halide hand print
11⅜ × 14¼ inches

Untitled (Figure #12),
2000
Zimbabwe
Silver halide hand print
10¼ × 13⅛ inches

Untitled (Figure #26), 2010
Rwanda
Silver halide hand print
10¼ × 13⅛ inches

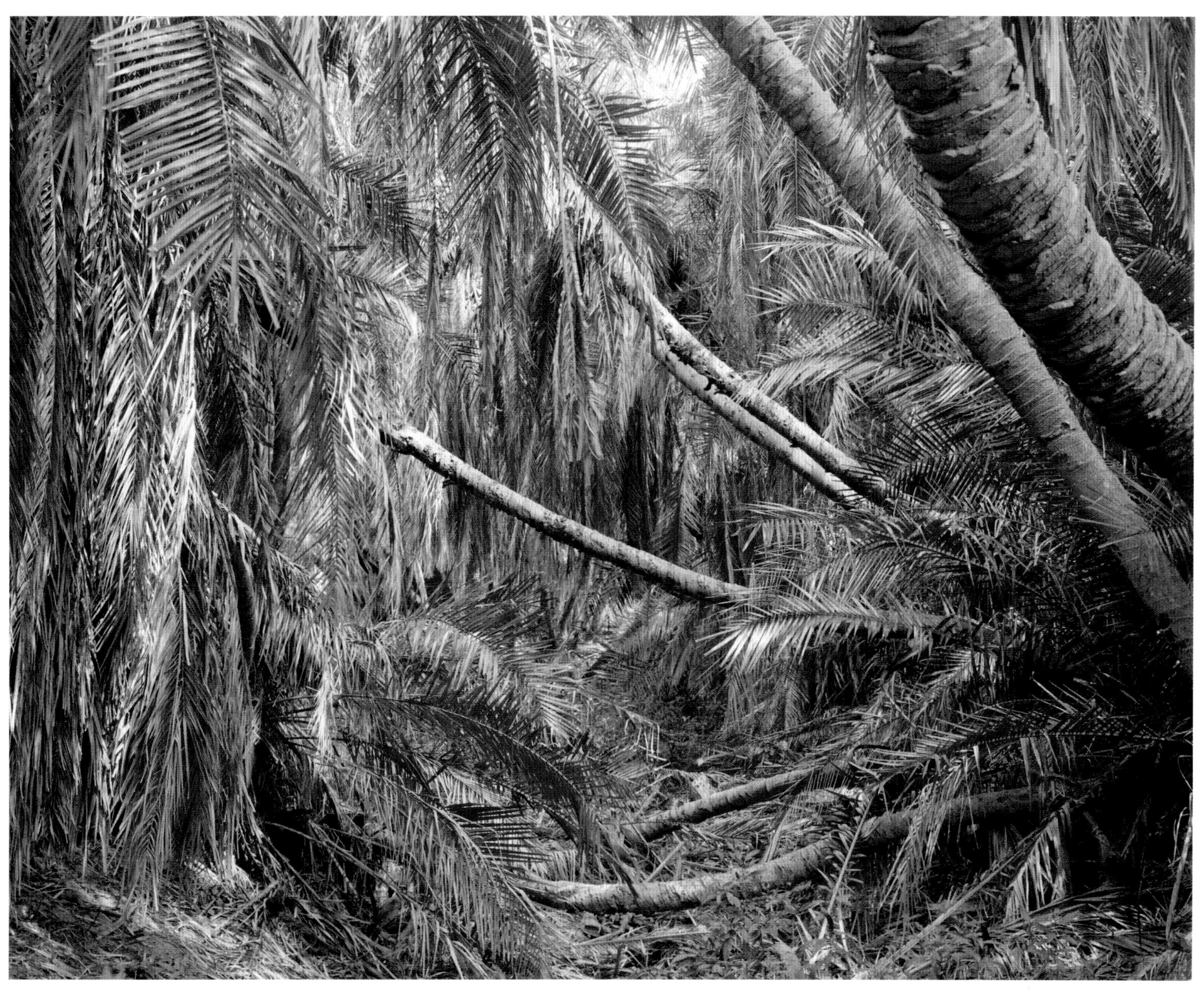

 Lisa Aronson

art in their "all-over" compositional play of branches, roots, and tangled undergrowth.[7] But Meyer's images of the African forest, featuring a level of visual detail that overwhelms the eye, intend to engage us in the beauty of an aspect of Africa's landscape seriously threatened by deforestation.

Papageorge's imposing and poetic landscape performance project, *Africa Rifting: Lines of Fire, Namibia/Brazil*, completed in 2001 and documented in *Environment and Object* by the artist's photographs and video, addresses landscape in a manner akin to the work of Earth artists, such as Robert Smithson, Michael Heizer, Richard Long, and Christo and Jeanne-Claude, which similarly transforms, wraps, or otherwise uses the earth as a canvas for ideas. Like them, Papageorge creates on a grand scale by stretching bright red banners across the beaches of Namibia in southern Africa and Brazil in South America to produce a visual commentary on Africa's ecological history.

The two points Papageorge draped with red banners mark a geological rift that opened about 135 million years ago, when the supercontinent Gondwanaland split in two, creating

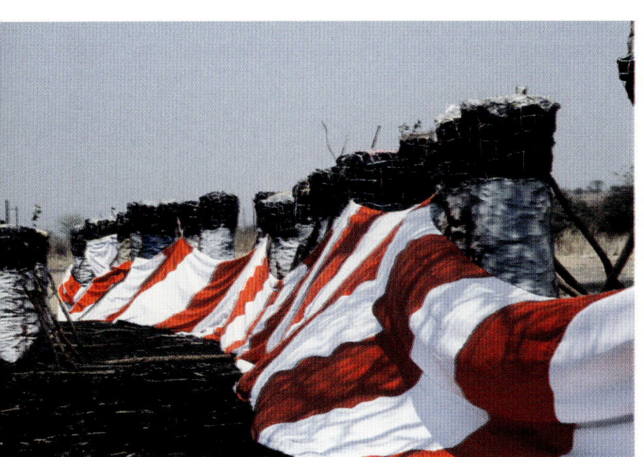

two smaller continents, South America and Africa. The rift's enormous seismic activity resulted in the creation of lakes and mountain ranges, some volcanic, along the present Rift Valley region of East Africa. The most noteworthy volcanic activity has occurred at Mount Kilimanjaro, Africa's highest mountain. The legacy of Gondwanaland has human consequences as well. The ice caps along Kilimanjaro's three volcanic cones are now threatened by deforestation and the illegal sale of charcoal derived from that wood. Papageorge highlights that black market by draping a red-and-white cloth banner along the front of the charcoal bundles being sold to warn of the impact that deforestation can have on local wildlife.

Africa Rifting comments symbolically on the geological, ecological, political, and ultimately religious relationships that connect people to the land and to each other, but that also divide them. That the second half of her landscape performance took place on September 15, 2001, only four days after the 9/11 attacks, added an additional layer to the themes of rift, connection, and division.

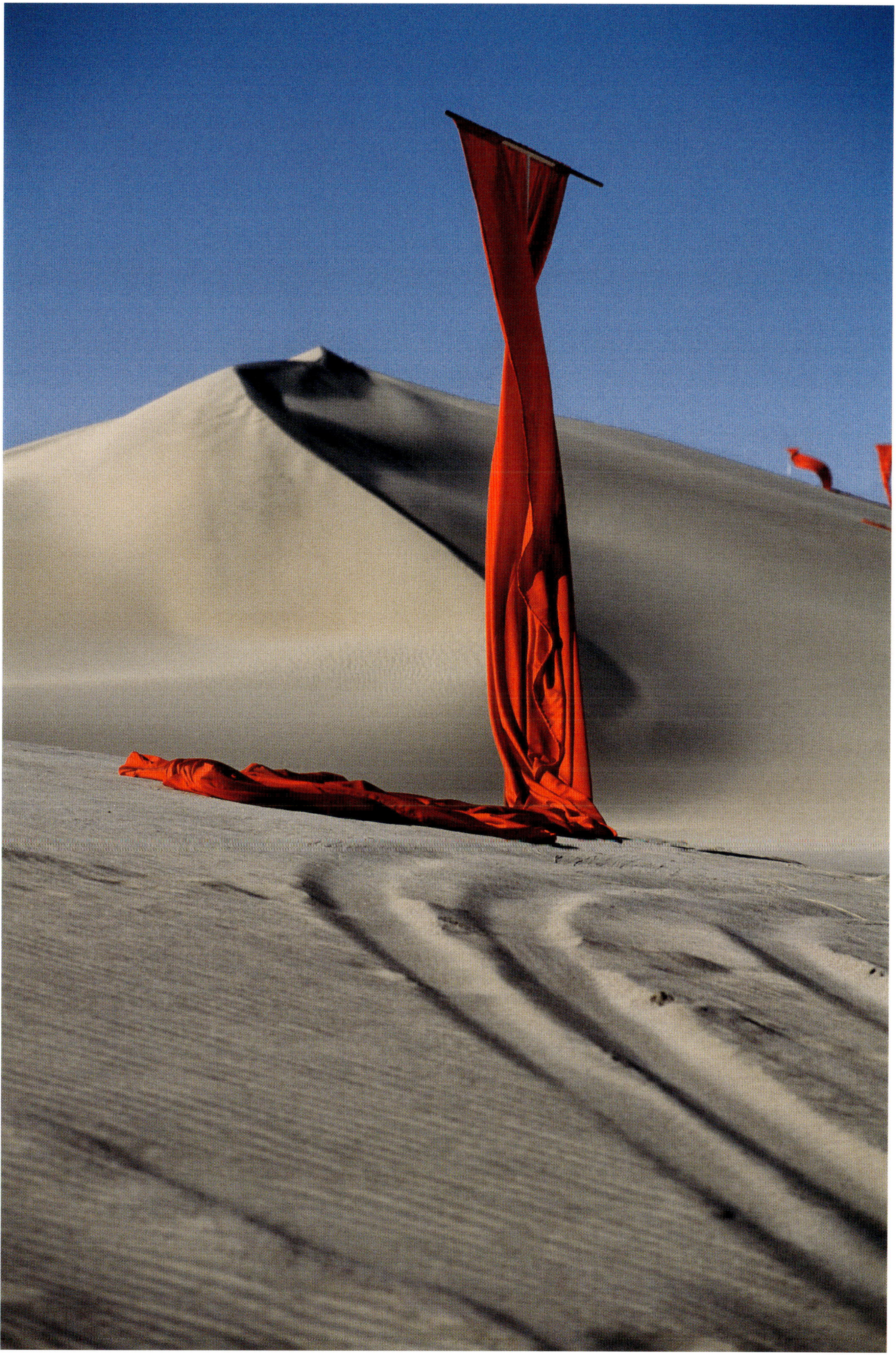

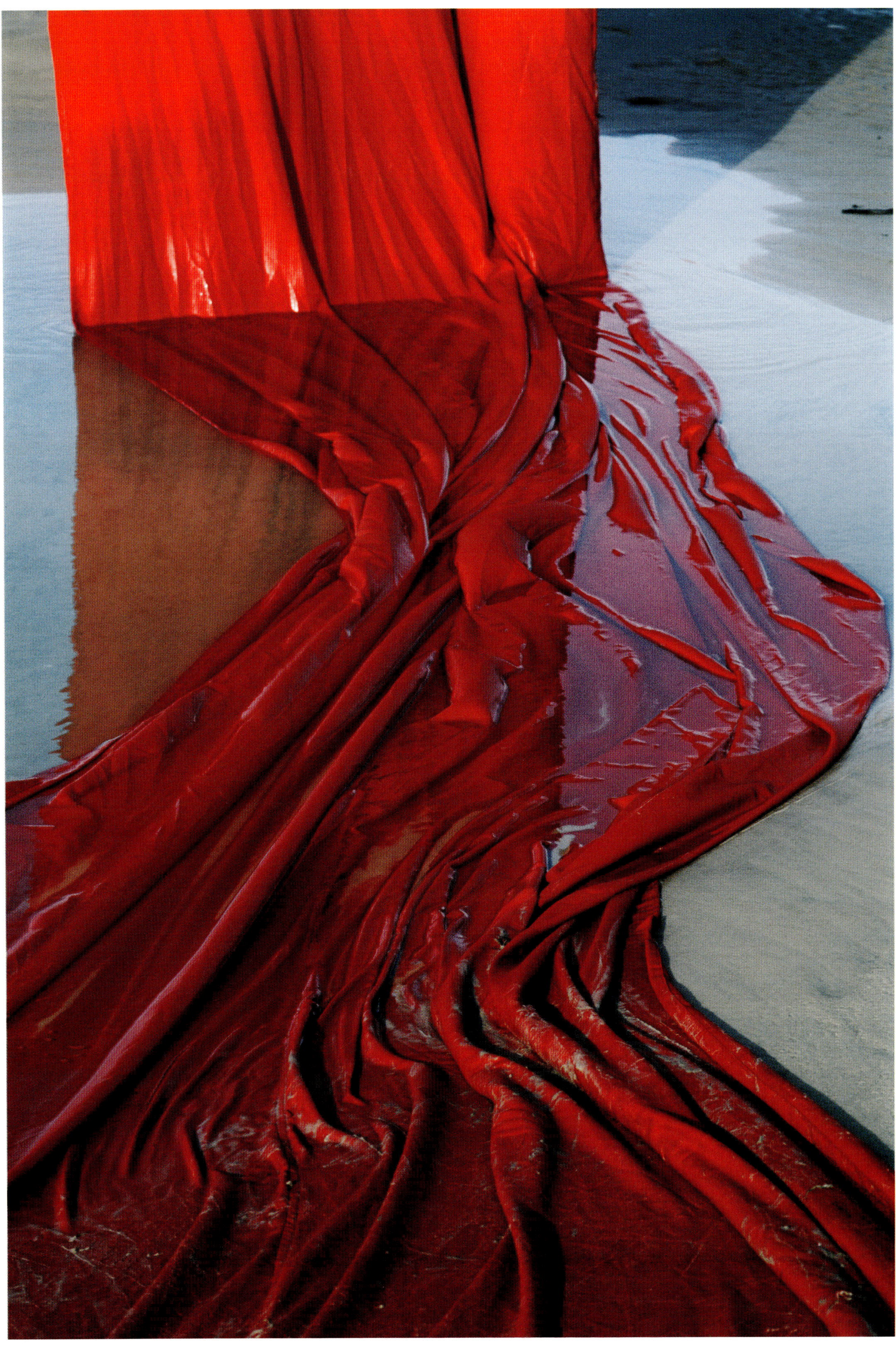

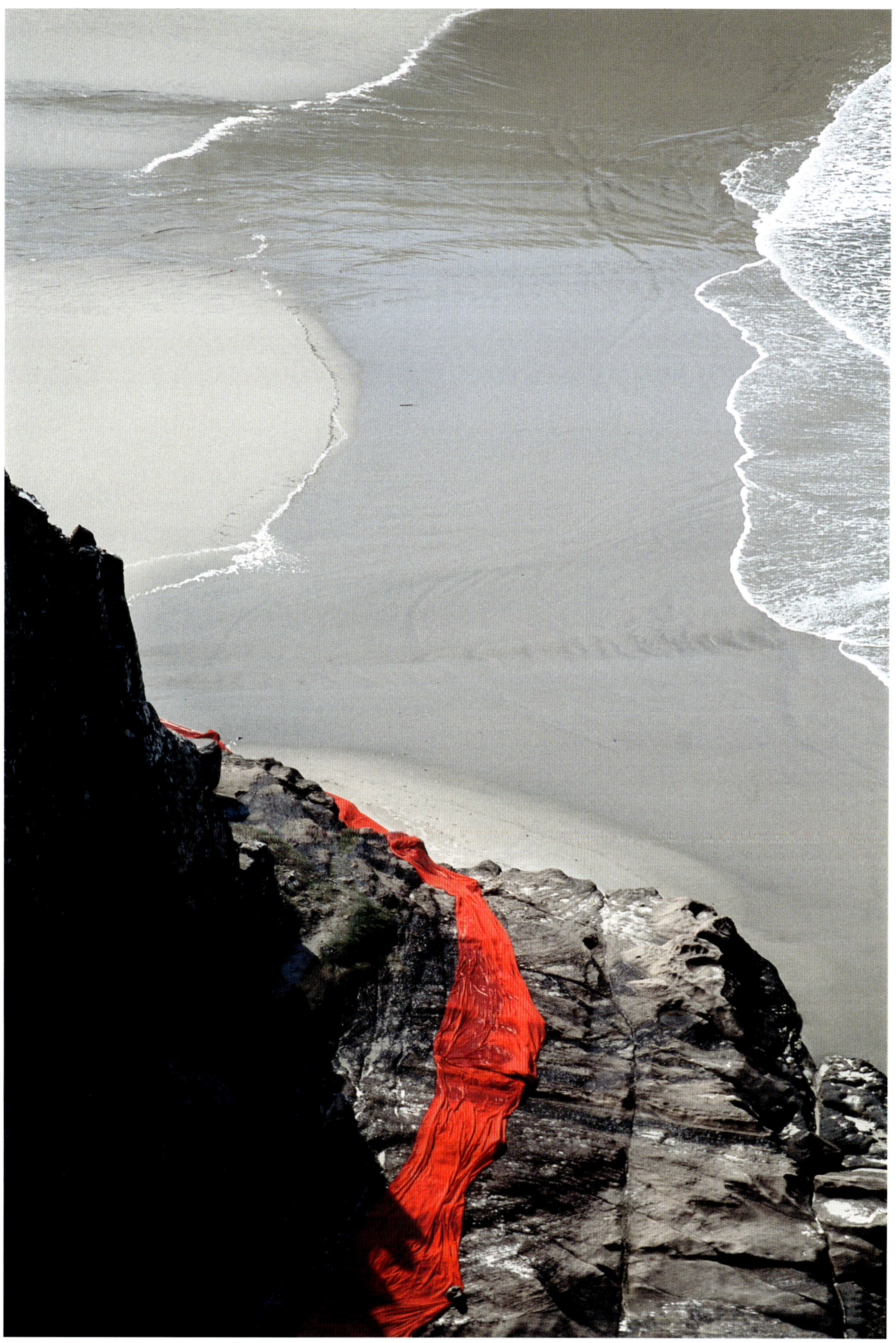

GEORGE OSODI
Oil Spill Near Farm Land
Ogoni, 2007
Digital inkjet print
31¾ × 47½ inches

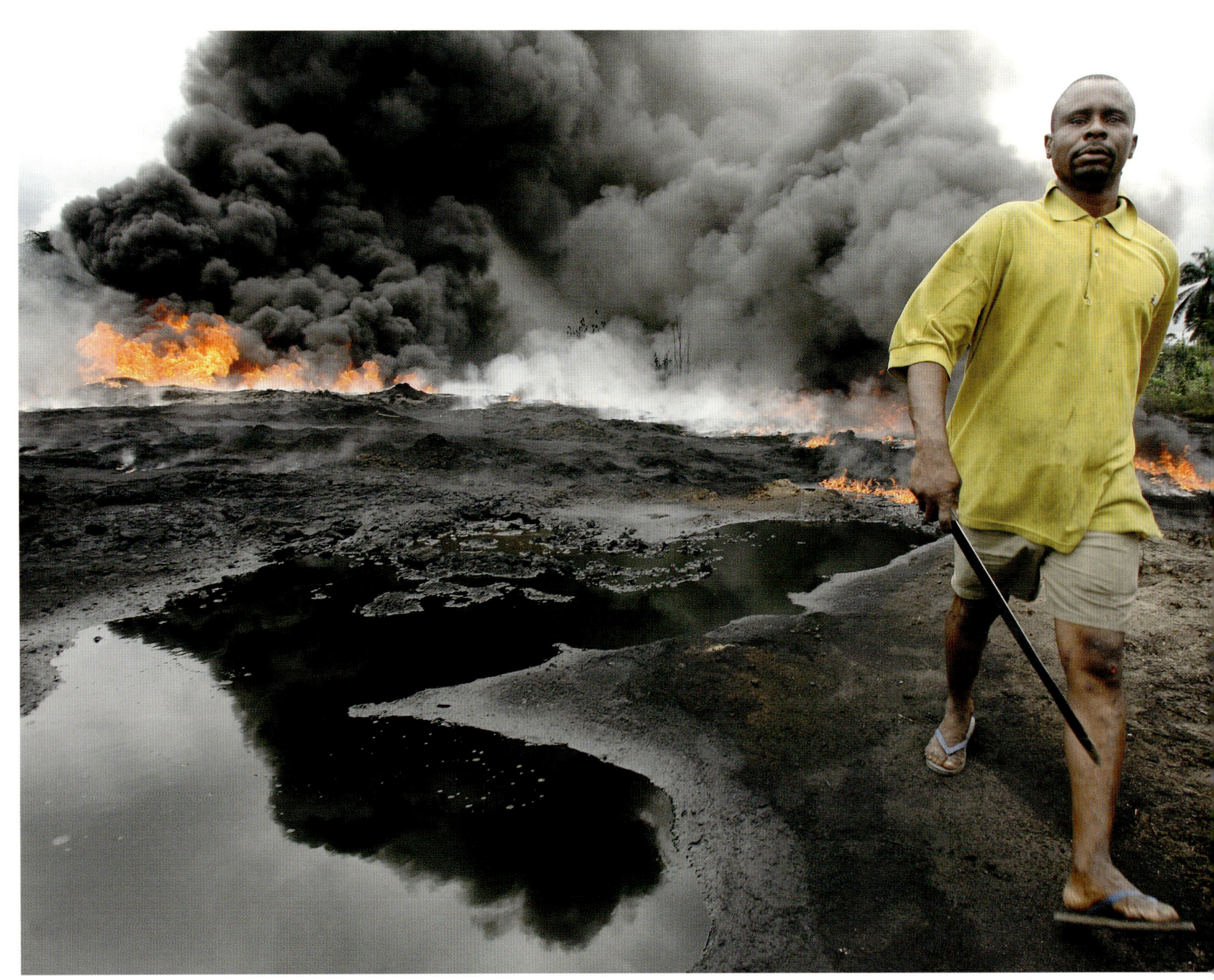

Environment through the Photographic Lens

Africans and Europeans began photographing the African landscape from the time photography was first introduced in the early nineteenth century. Europeans frequently took or commissioned photographs to satisfy their own interests in the African landscape. Photography proved useful for natural scientists, who discovered a treasure trove of new zoological and botanical specimens and helped amass significant collections in natural history museums.[8] Colonizers and merchants commissioned photographs of the African terrain in part because they saw in it a wealth of natural resources, including palm oil, rubber, gold, and copper, that could greatly benefit their own industries and economies. African photographers quickly accommodated, and even exoticized and otherwise enhanced the landscape imagery to appeal to the sensibilities of their European clients.

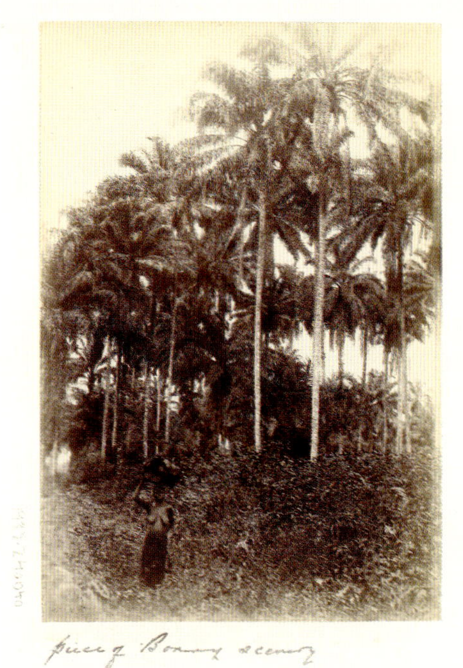

The works by Nigerian photographer **George Osodi**, taken a century later, challenge and provoke rather than please the viewer. They make explosive statements about the exploitation of crude oil, now the dominant resource in the Niger Delta. Osodi's photograph *Oil Spill Near Farm Land Ogoni* (2007) depicts a young man at the right moving briskly into our space, away from the burning oil wells and oil-polluted waters that surround him. Osodi's disturbing imagery echoes the delta inhabitants' voices protesting against such companies as Mobil and Exxon, whom they blame for destroying their environment and stealing their resources. The reference to Ogoni in the title links it with Ogoni environmental activist and writer Ken Saro-Wiwa, executed with others in 1995 for protesting the pollution that multinational oil companies had brought to this region of the delta. Saro-Wiwa has since become a culture hero in the eastern Niger Delta, and Osodi's photographs stand as chilling reminders of the circumstances leading to his death.

     Lisa Aronson

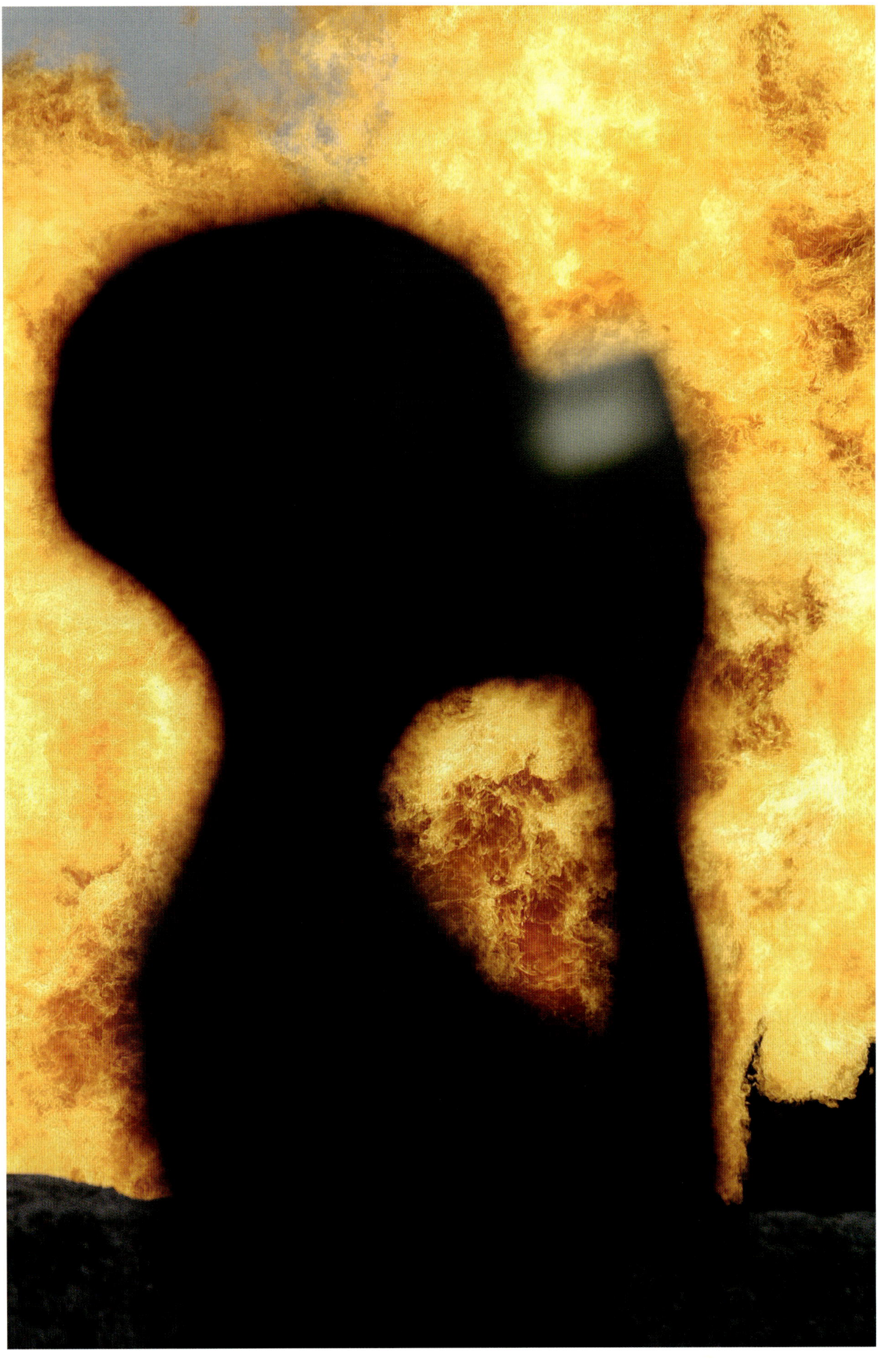

LEFT TO RIGHT
GEORGE OSODI
Niger Boy, 2006
Digital inkjet print
47⅝ × 31⅞ inches

Blackened Explosion, 2005
Digital inkjet print
47½ × 31¾ inches

   Land, Landscape, and Contested Spaces

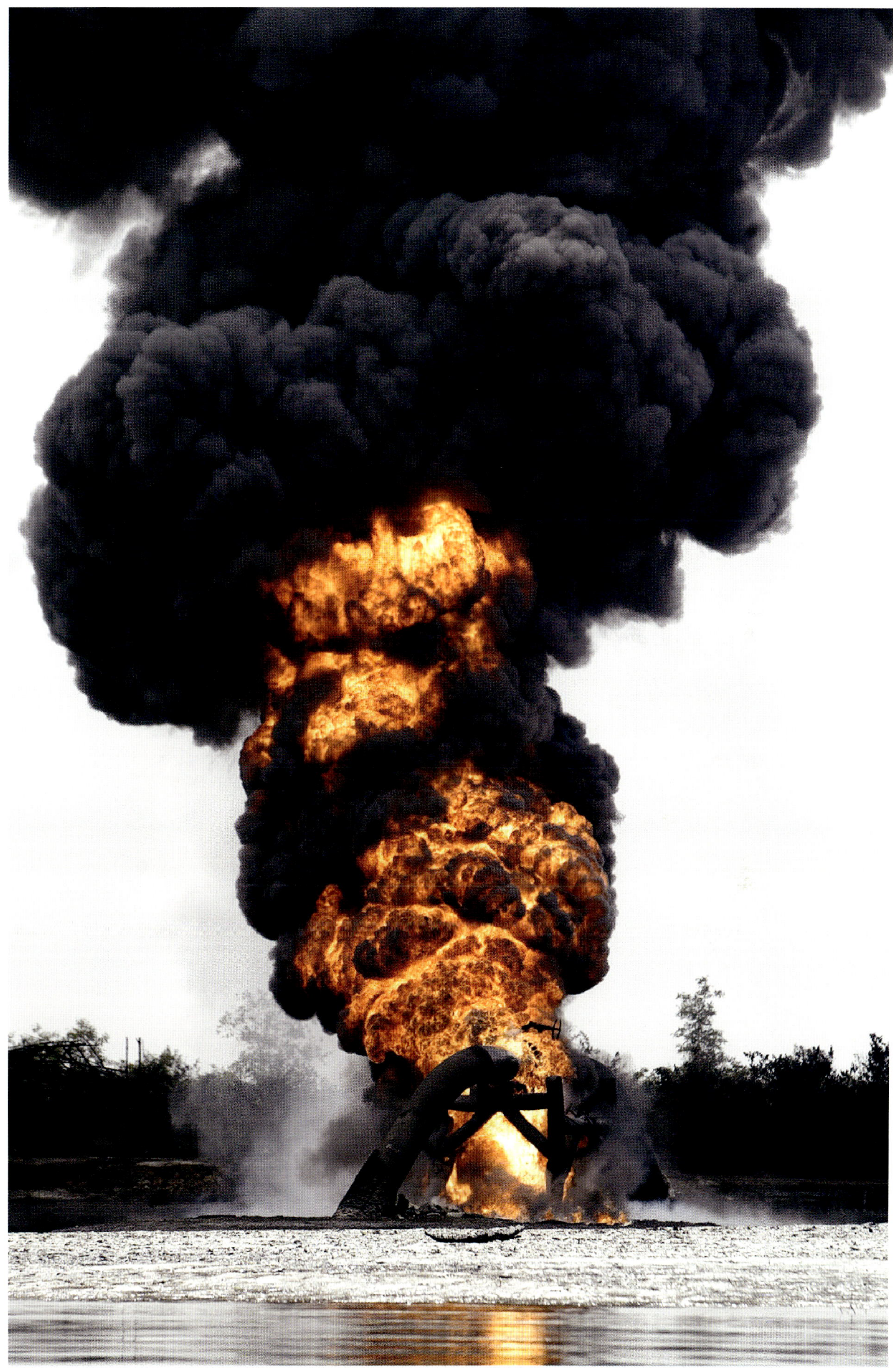

Lisa Aronson

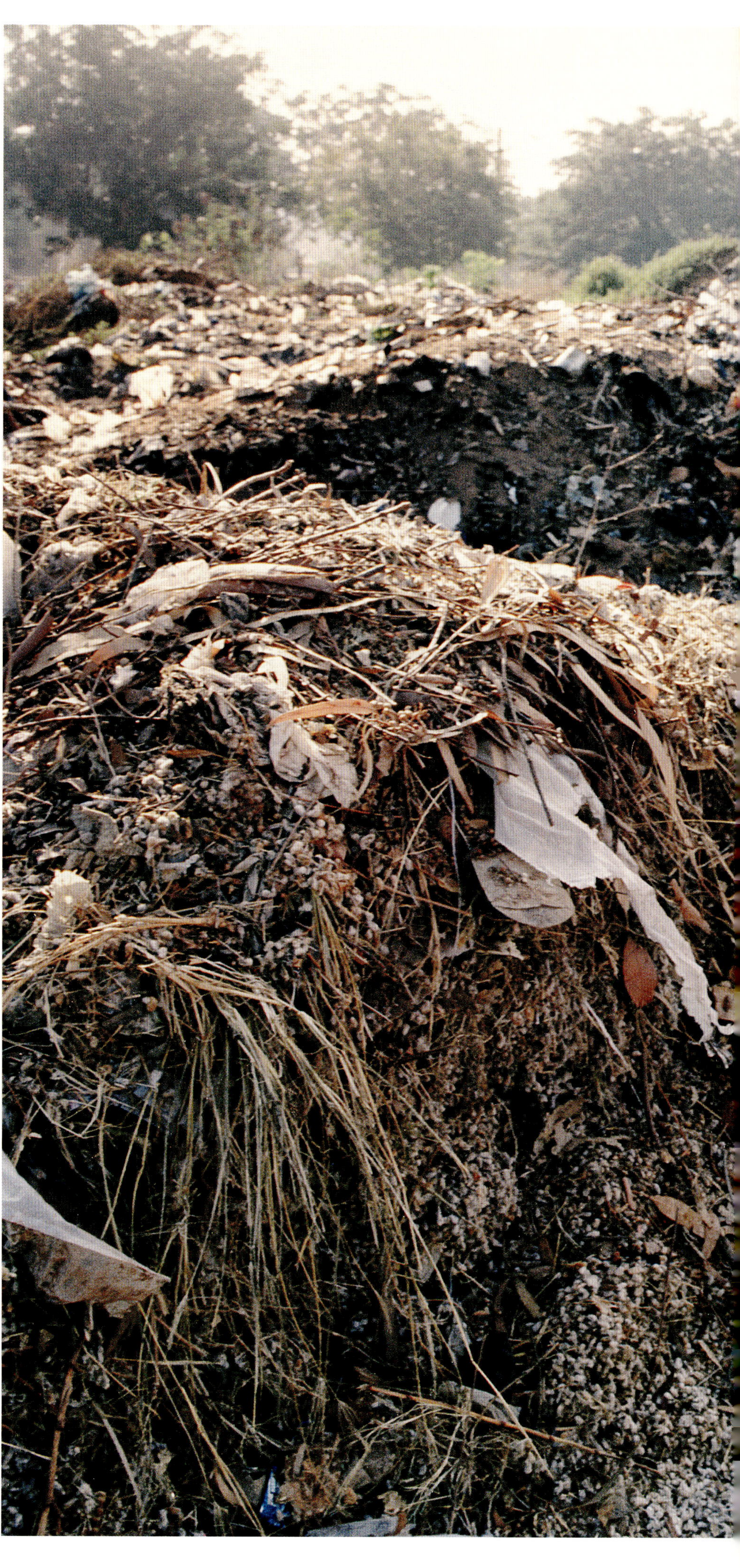

ZWELETHU MTHETHWA
Untitled (from the
**Contemporary Gladiators
Series**), 2008
Chromogenic print
60⅝ × 78¼ inches

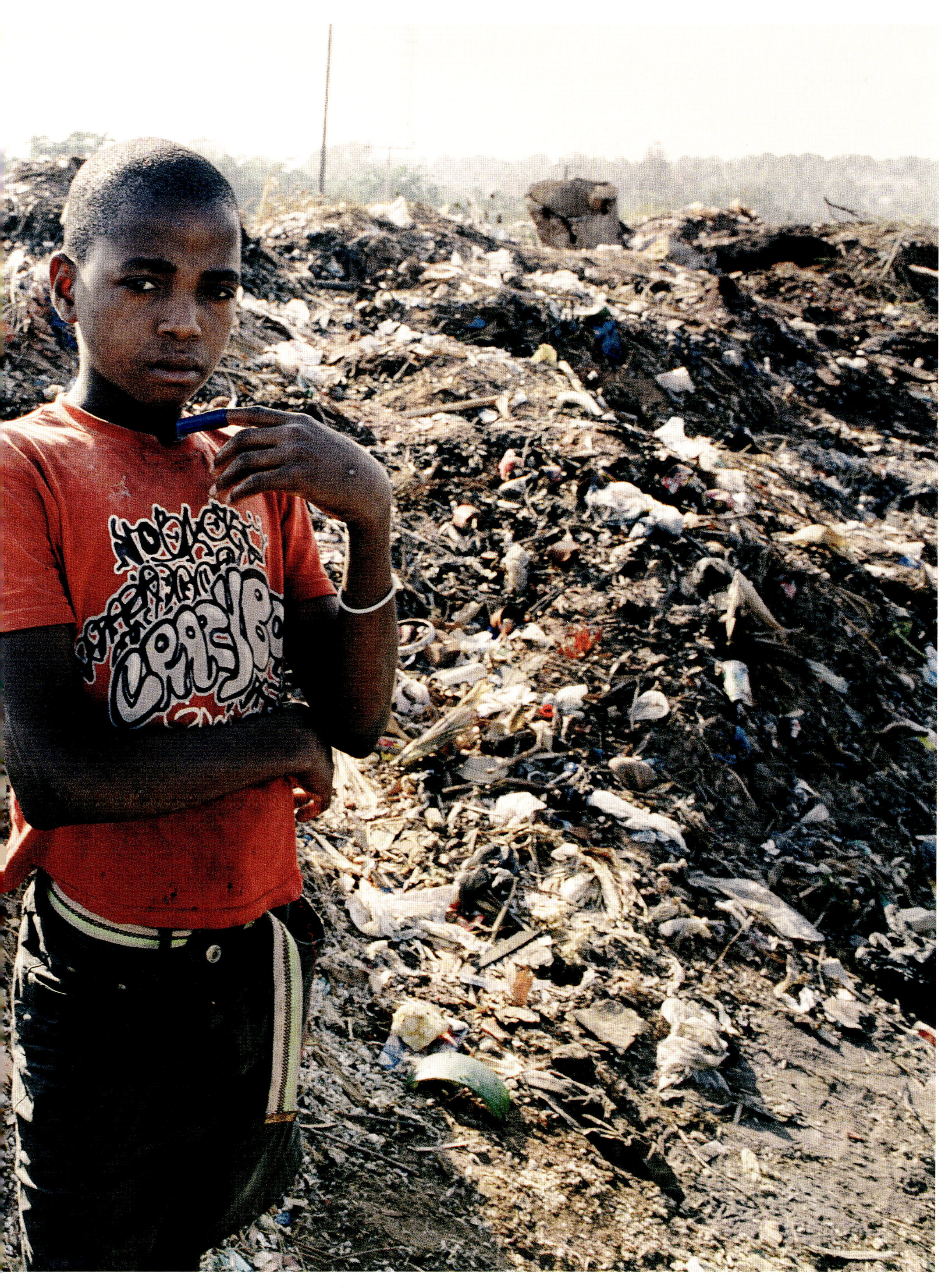

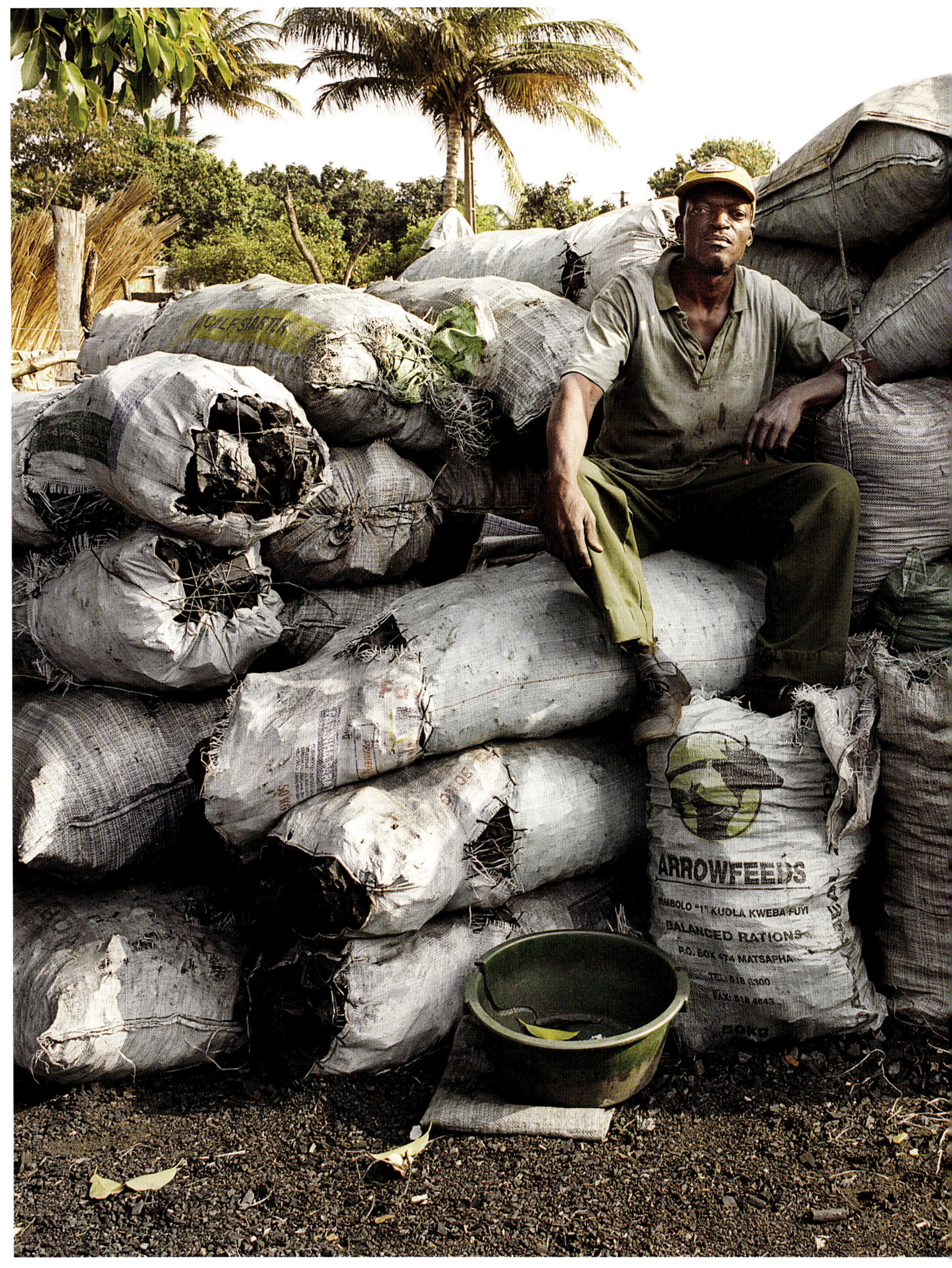
WILFSARTE
ARROWFEEDS
IMBOLO "1" KUDLA KWEBA FUYI
BALANCED RATIONS
P.O. BOX 474 MATSAPHA
TEL: 518 0300
FAX: 518 4643
50kg

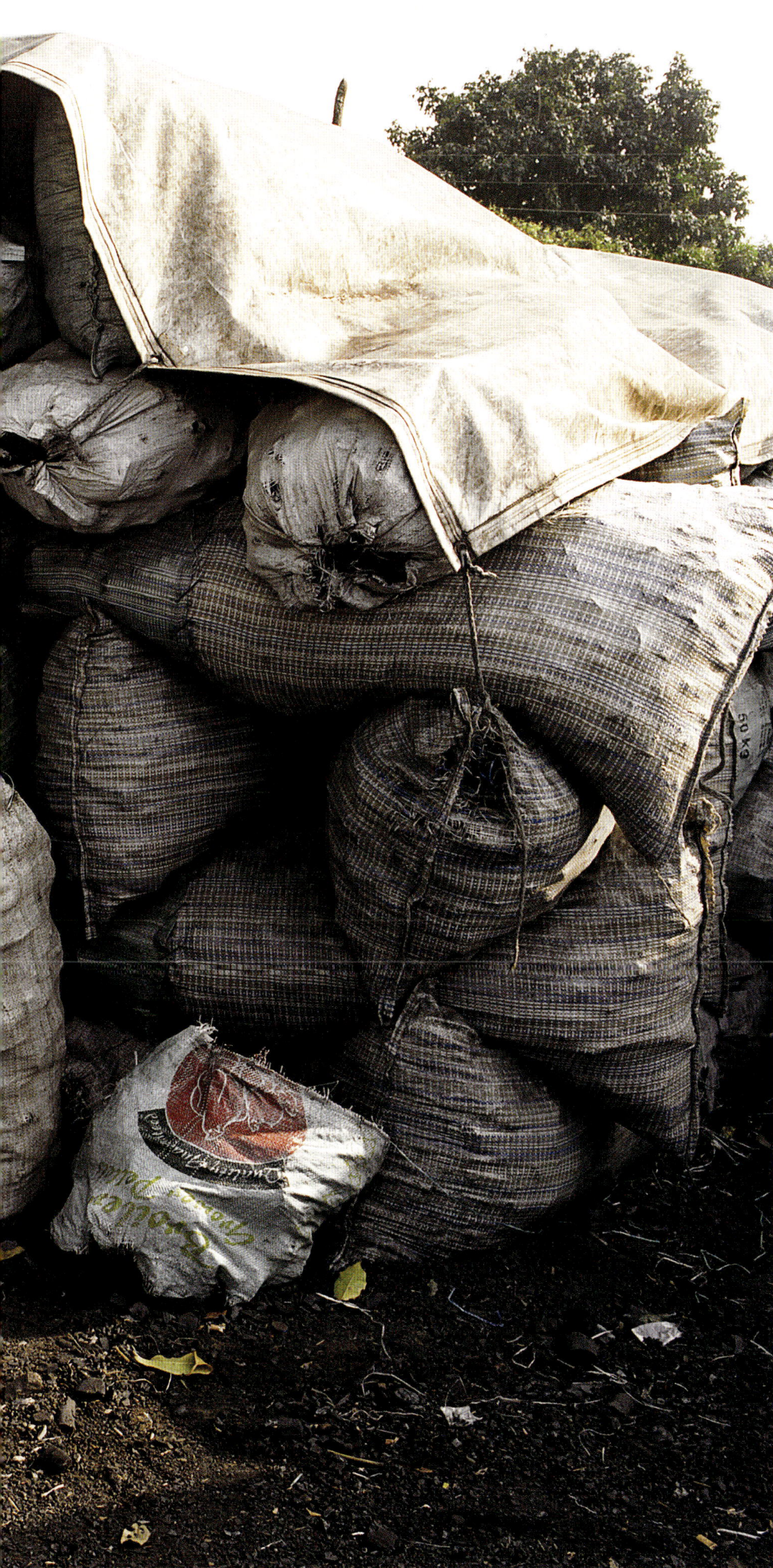

ZWELETHU MTHETHWA
Untitled (from the
Coal Miner Series), 2008
Chromogenic print
61½ × 79⅛ inches

Zwelethu Mthethwa's portrait photographs belong to the genre of environmental portraiture, as opposed to studio portraiture. He captures people in their home or work environment to shed light on their life experience, emphasizing the dignity and character of these black South Africans. Each image captures the viewer's gaze, which looks out pointedly and directly. Mthethwa's two portraits in *Environment and Object*, one from his *Contemporary Gladiator Series* and the other from his *Coal Miner Series*, offer poignant commentary on the human investment in land and its resources, and on related questions of race, class, and identity in South Africa. The image from the *Gladiator Series* features a child standing amid the trash pile he was surely rummaging through for anything of worth. The *Coal Miner Series* image reveals the harsh conditions of migratory laborers who come from other parts of Africa—Mozambique, in particular—to South Africa, where land is government-owned and monitored, giving workers a reliable though minimal income. In each image, the landscape functions as more than a backdrop to the people; it is part and parcel of their displaced lives and identities.

Rather than employ black and white, which, according to Mthethwa, evokes the journalistic photography associated with Apartheid, his large-scale, richly patterned photographs capture his subjects' dignity through color and formal composition. For example, the graffiti-like word "crazyboy" written across the young boy's T-shirt mimics the surface design of the pile of human detritus behind him, and the occasional specks of red in that pile match the red in his shirt. Likewise, the green of the plastic bucket, occasional woven polypropylene bags, and background foliage in the *Coal Miner* photo coor-dinates with the worker's pants. Yet Mthethwa leaves his photographs untitled, as always, and his subjects anonymous to "spark further inquiry in terms of who and what is pictured in the actual portrait,"[9] as he notes.

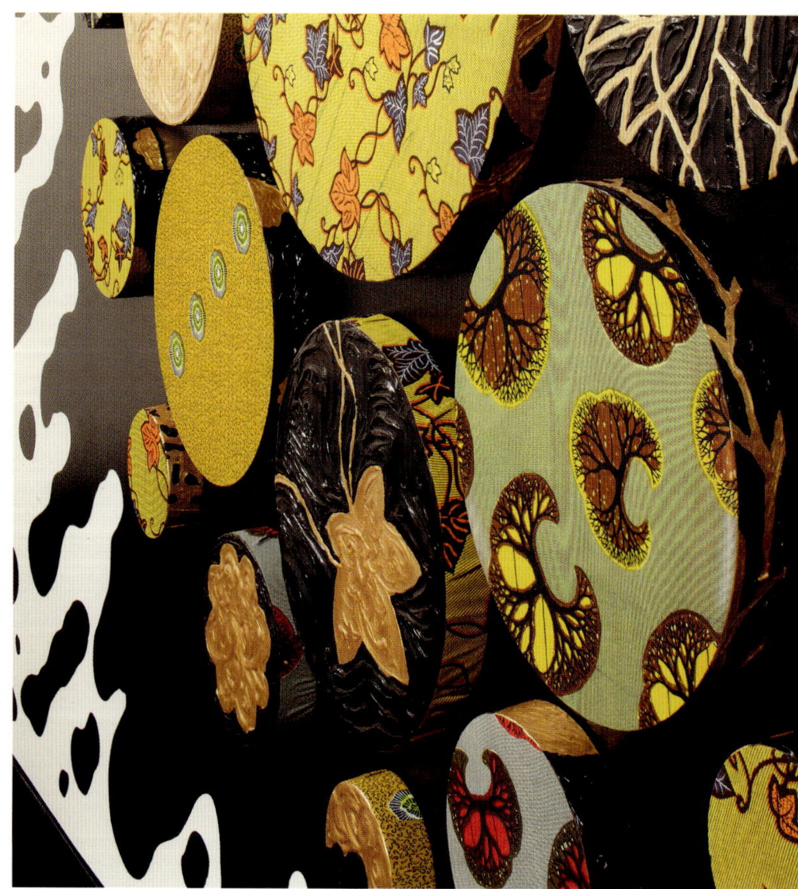

Expanding Views of Landscape

Two other Nigerian artists, **Yinka Shonibare** and **Sokari Douglas Camp**, would hardly consider themselves environmentalists, yet both feel compelled to create art that addresses the seriousness of Nigeria's oil problems, using their own highly individual palettes. Shonibare admits that he always tries in his work to challenge stereotypes

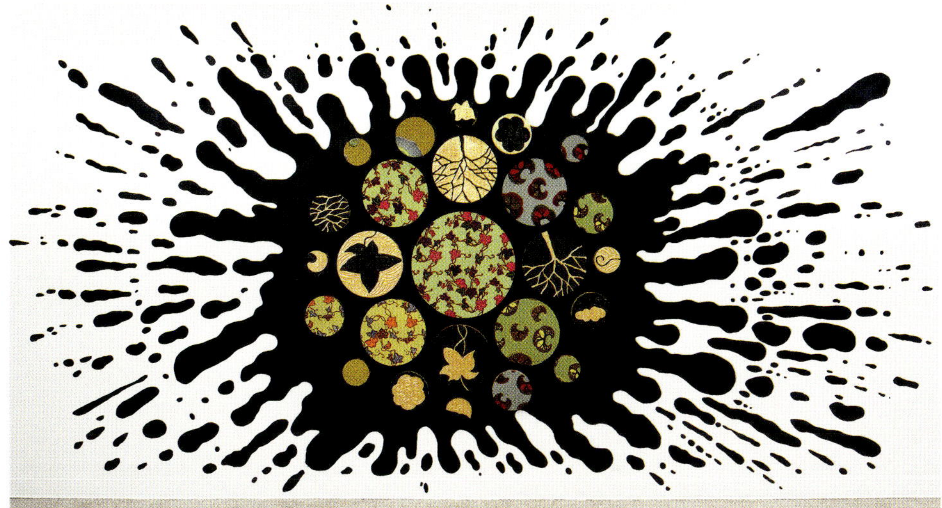

and assumptions about Africa, as well as critique Europe's ongoing relationship with it. He focuses these issues by using colorful, richly patterned, factory-printed textiles that have a long history in Africa. Originally imported from Indonesia to Africa through trade, and subsequently manufactured in Holland, England, and most recently Africa itself, these factory cloths now dominate the textile stalls of any African market. Seeing these fabrics as symbols of globalization, he uses them to fashion the Victorian-style attire worn by his headless mannequins or to add color and pattern to his flat surfaces, in each case reminding us of the impact of Victorian English involvement with Africa and, in turn, how closely connected Africans have long been to the rest of the world.

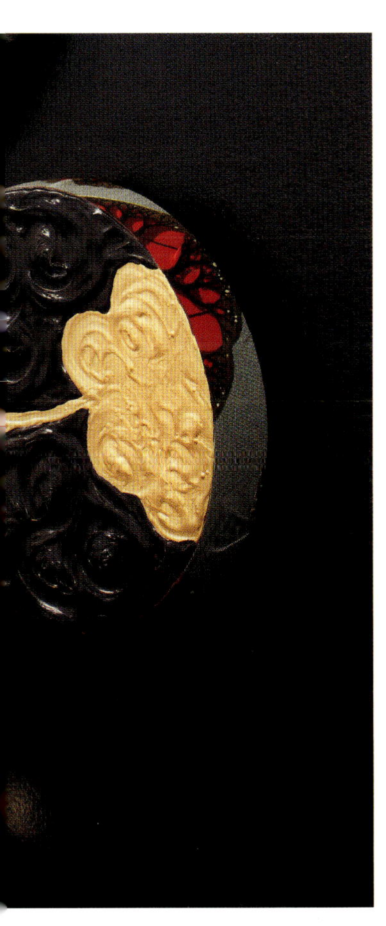

Factory prints constitute a vital component of Shinobare's two wall paintings *Black Gold I* and *Black Gold II* (both 2006), which emphasize the power and destruction that oil has imposed on the Niger Delta's environment. These paintings are dominated by a large exploding splatter of black acrylic painted directly on the wall. In its center, Shonibare attaches round canvases covered in factory floral prints with imagery suggesting nature and sometimes painted. Spectacular in scale and nearly identical in appearance, both versions of *Black Gold* convey oil pollution's serious encroachment on the environment and implicitly indict the power of the multinational oil companies who are responsible. The painting invites a reading as a bird's-eye view of the Niger Delta, its numerous small islands now surrounded by oil-polluted water.

Shonibare followed these two wall works with a series of environment-related images combining drawings and collage. At first glance, they display a lyrical ménage of drawn and collaged flowers that play with abstract and representational shapes. But a closer look

Lisa Aronson

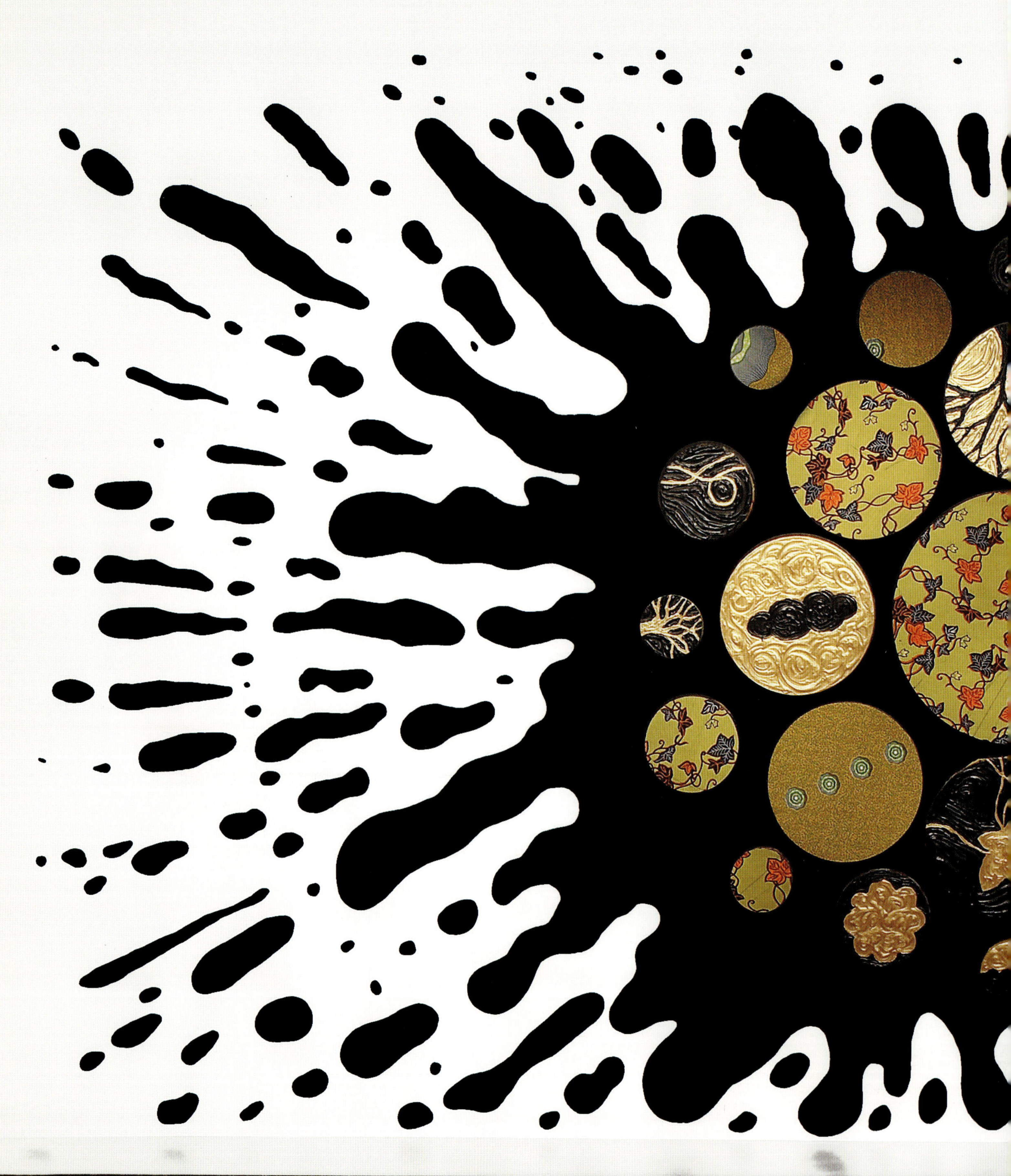

     Land, Landscape, and Contested Spaces

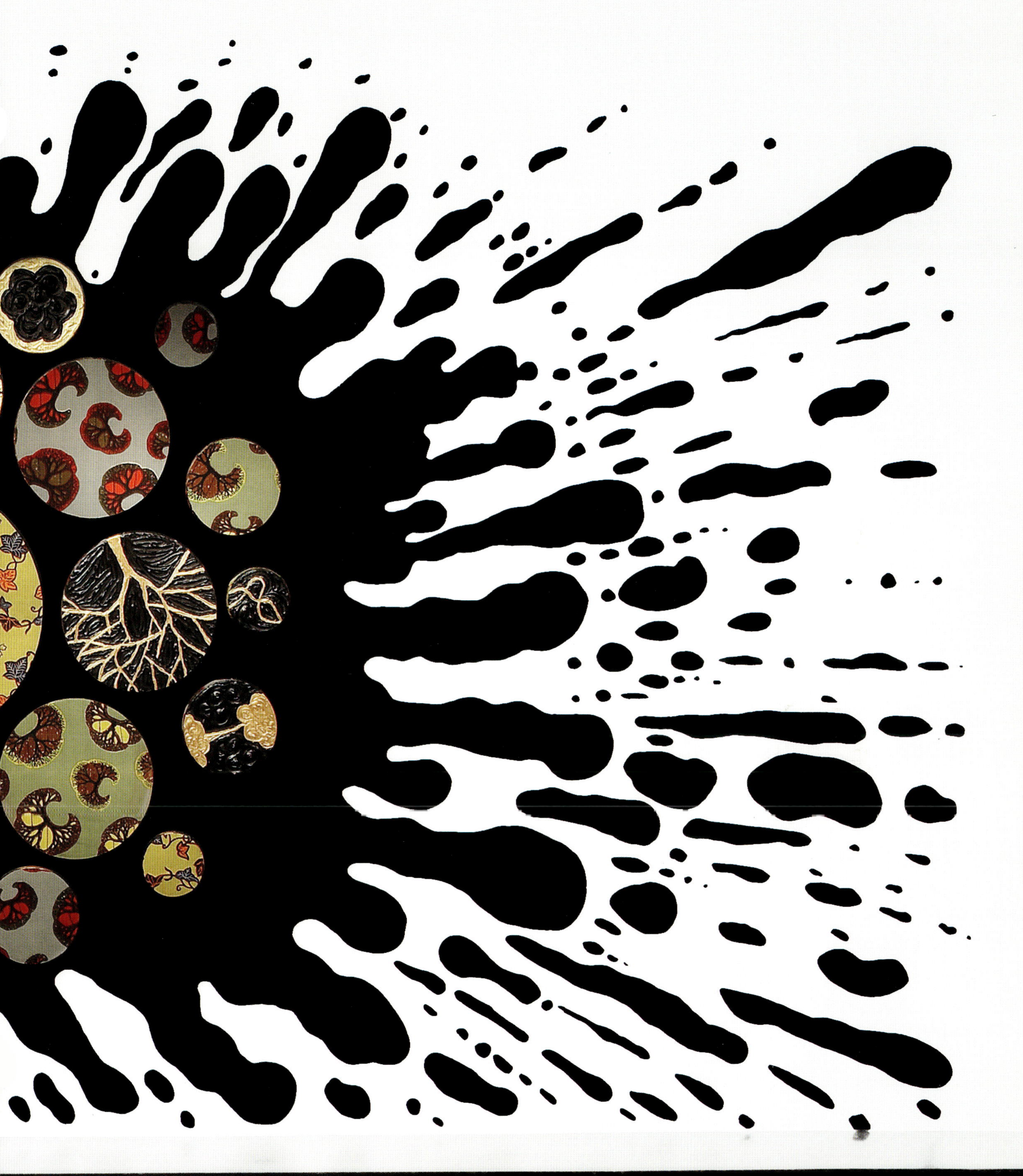

Lisa Aronson

reveals disturbing, environmentally related imagery. One of the flower-shaped cutouts in a work titled *Climate Shit Drawing 1* (2008) comes from a newspaper clipping with the headline, "Rising Cost of Oil," giving these images, like the *Black Gold* pieces, a political and economic context.

Unlike Shonibare, whose family roots stem from the Yoruba region to the west, sculptor Sokari Douglas Camp is a Kalabari Ijo woman born and bred in the Niger Delta. She left the area at twenty-one to study in Southern California and then London, where she has remained ever since. The two works by Douglas Camp in the exhibition represent a marked departure from her earlier work, *Spirits in Steel, The Art of the Kalabari Masquerade.* This signature work consists of life-sized, welded figures representing the whole cast of characters in a typical Kalabari masquerade: the masqueraders themselves, the drummers, and the spectators on land and in an elaborately decorated regatta. Douglas Camp used this kinetic, playful, and uplifting sculptural installation to reconnect with her Kalabari roots.

Though similar in their construction and scale, Douglas Camp's two sculptures in this show consider the more sinister side of the Niger Delta, again by addressing the oil conflict. Her sculpture *Assessment* (1998) offers a perspective on the delta landscape, using imagery instantly recognizable to its people. It features someone holding up a photograph as during a funerary procession to honor the deceased. However, in this case, the memorial image depicts an oil refinery of the kind that now dominates the Niger Delta coastline. A mixture of oil and water poured into the bottom of the two pieces of glass that sandwich the image comments on the unhappy marriage of those two materials. By depicting such imagery in a format evocative of funeral processions, she reminds the viewer of the destruction and death caused by the oil conflicts in the area.

    Land, Landscape, and Contested Spaces

YINKA SHONIBARE
Climate Shit Drawing 1, 2008
Lithographic print with
silkscreen glaze, fabric, foil,
and die-cut Somerset
radium white satin paper
Edition of 200, 13½ ×
20 inches

Funeral Procession,
Akwete, Nigeria, 1978

SOKARI DOUGLAS CAMP
Assessment, 1998
Steel, glass, acetate, and oil
95¼ × 44 × 27⅛ inches

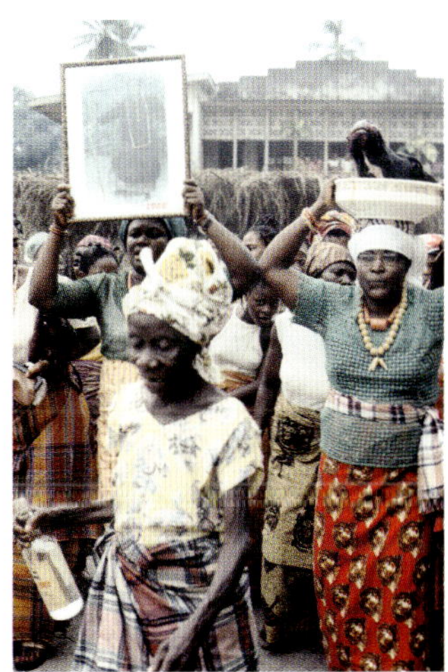

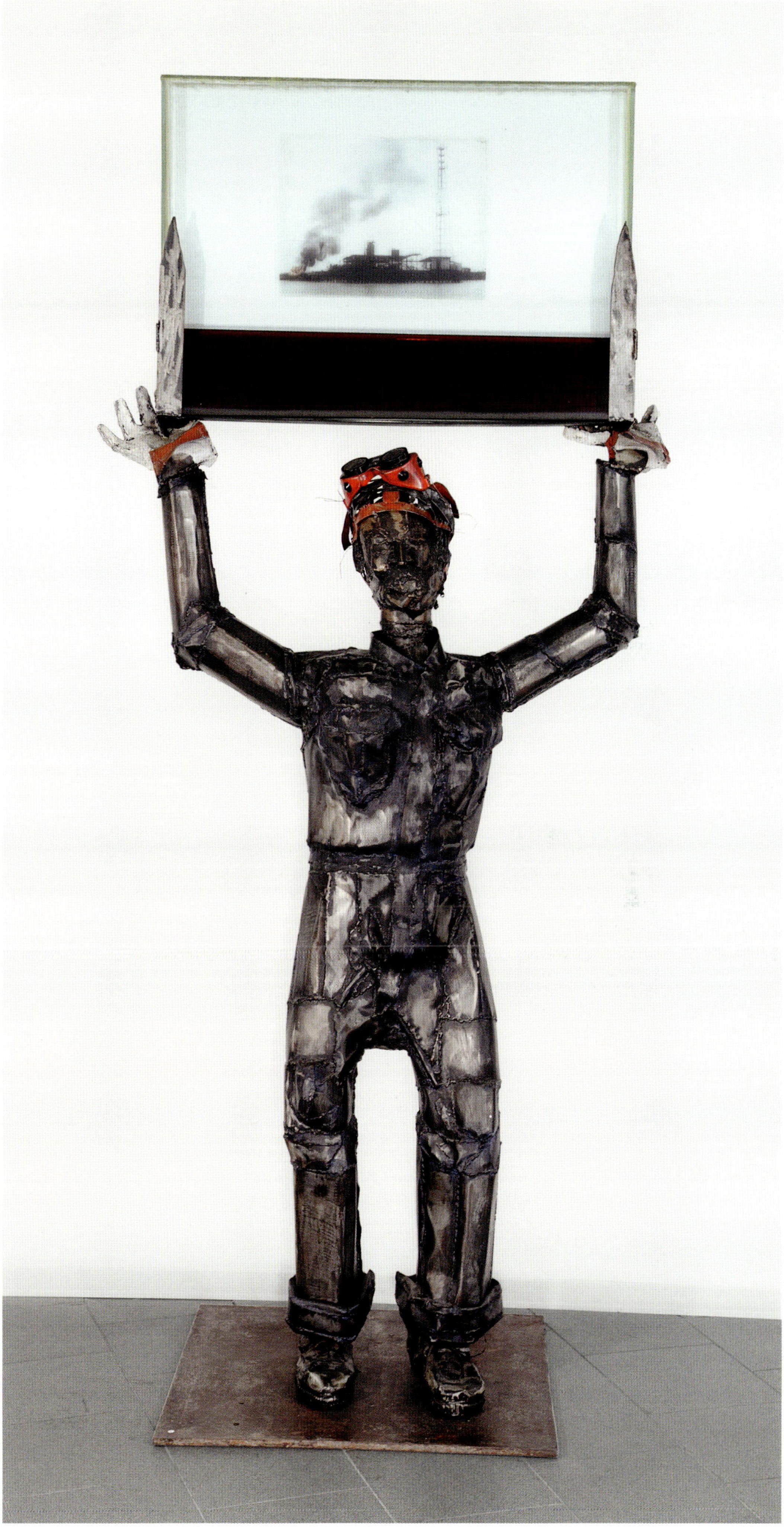

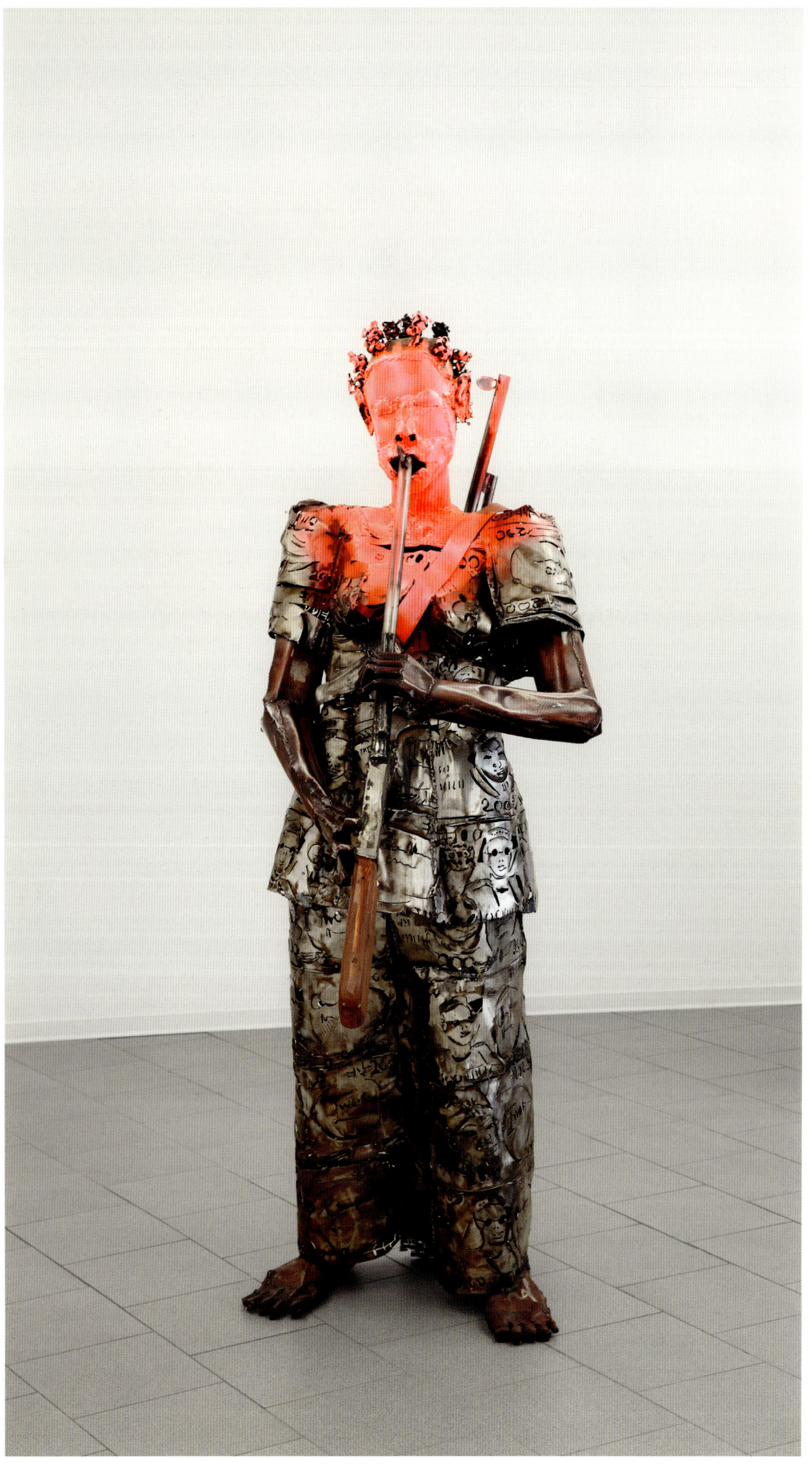

Teasing Suicide (2004) comments much less subtly on the human devastation caused by the oil crisis. Any internet search for images associated with the oil crisis in the Niger Delta immediately brings up photographs of armed militias. The largest of those, a group called MEND (Movement for the Emancipation of the Niger Delta) formed in 2006,

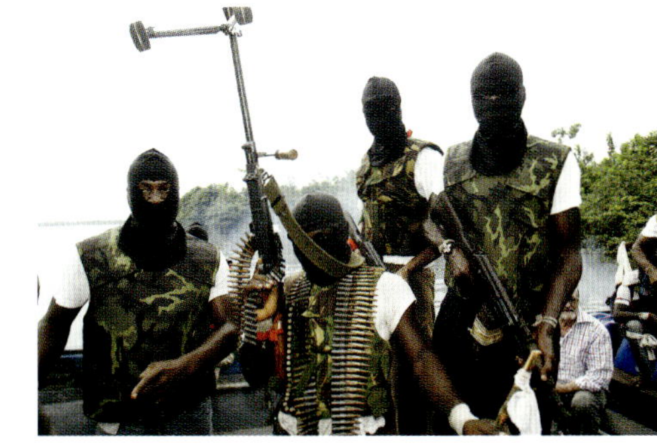

aims to expose exploitation and oppression in the Niger Delta, particularly with respect to petroleum operations. Widely circulated photographs show these militias holding assault rifles and other menacing-looking weapons, their individual identities hidden by cloth masks over their heads.

Women rarely join the militias, but Douglas Camp's *Teasing Suicide* uses the image of a female militant as a powerful tool in the African discourse on struggle and conflict.[10] Douglas Camp's sculpture of a woman pointing a Kalashnikov-type rifle into her mouth transforms the artist's militant into a martyr who turns the outward expression of violence and protest on herself, "teasing suicide" as the title suggests. Douglas Camp painted the head and neck of this surprising female figure bright pink, with the gun in her mouth, "to make people take notice."[11] She imprinted the woman's steel clothes with a pattern representing two different Nigerian banknotes, representing the south and the north. "It's what Nigeria is about," she explains. "The country is almost split because of the troubles in the Delta. So it seemed very appropriate that I chose these two icons from Nigerian politics: money and guns."[12]

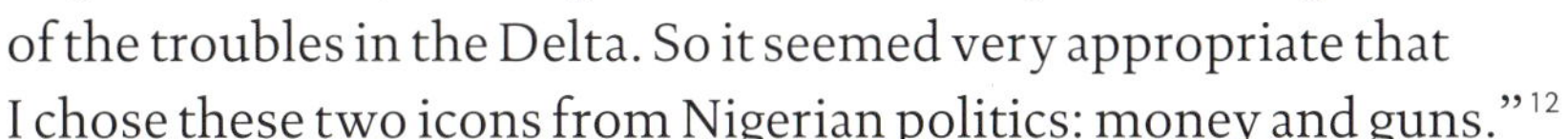

Douglas Camp's most imposing sculpture about the oil crisis is her tribute to the Niger Delta's culture hero Ken Saro-Wiwa, also alluded to in George Osodi's work.[13] She designed this prize-winning memorial in the form of a life-sized Nigerian-style bus, characterized by its top-heavy loads and words painted on the side (see page 134).[14] The top bears nine oil cans commemorating the nine protestors sent to the gallows on November 10, 1995; words on the bus's side, imprinted over a blood-stained banner, read, "I accuse the oil companies of practicing genocide against the Ogoni." Douglas Camp chose this unconventional form of memorial to satisfy the wishes of the Saro-Wiwa family, who preferred something more poignant and Nigerian-style than a Victorian-style monument. Douglas Camp also wanted her memorial to provide usable interior space for lectures and workshops, and to be portable enough to visit venues throughout Britain. Though it cannot be driven, the bus has appeared in multiple London locations, as well as in Bristol, Liverpool, and Birmingham, the cities that were most heavily involved in the slave trade.

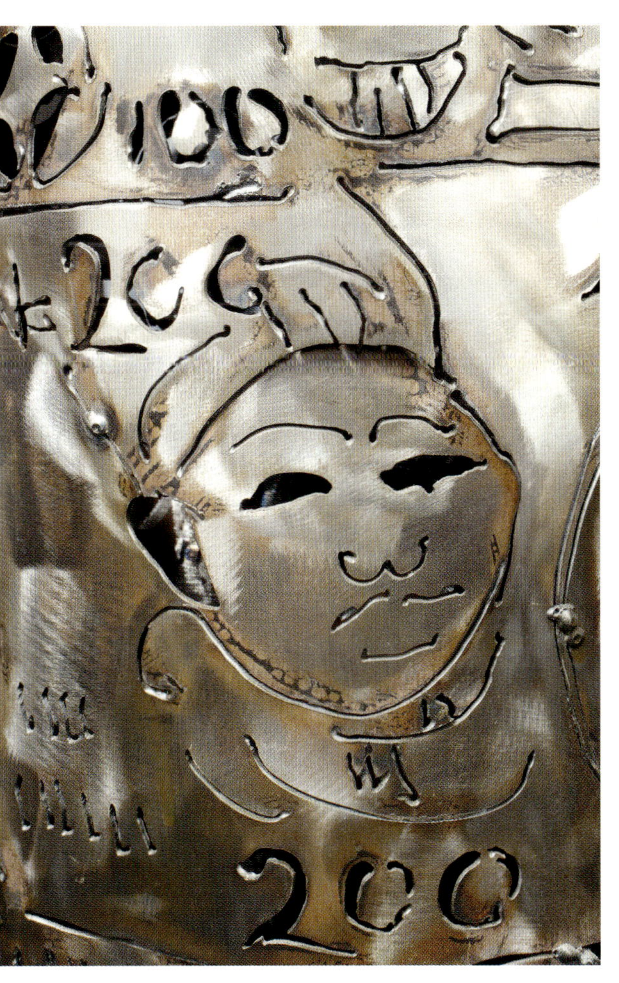

     Lisa Aronson

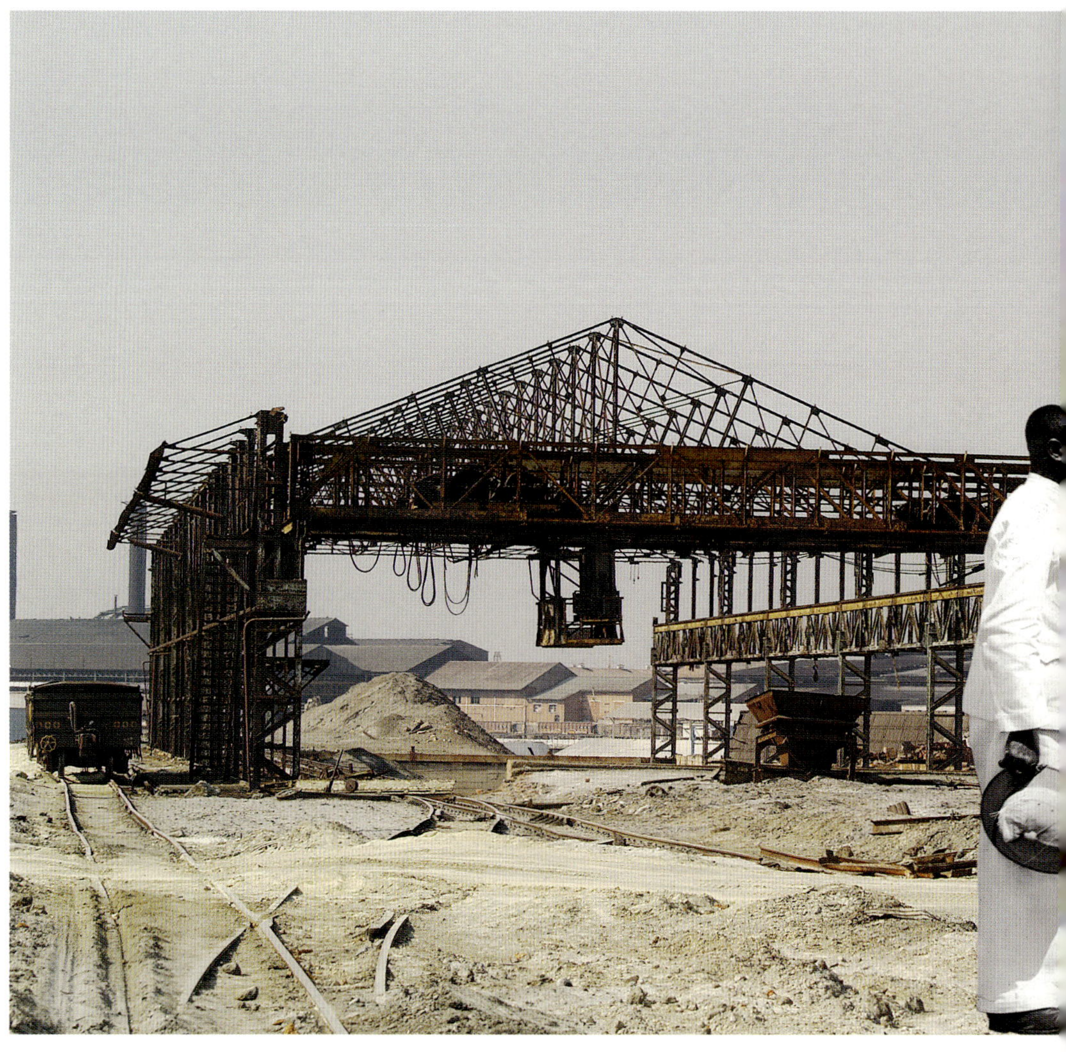

     Land, Landscape, and Contested Spaces

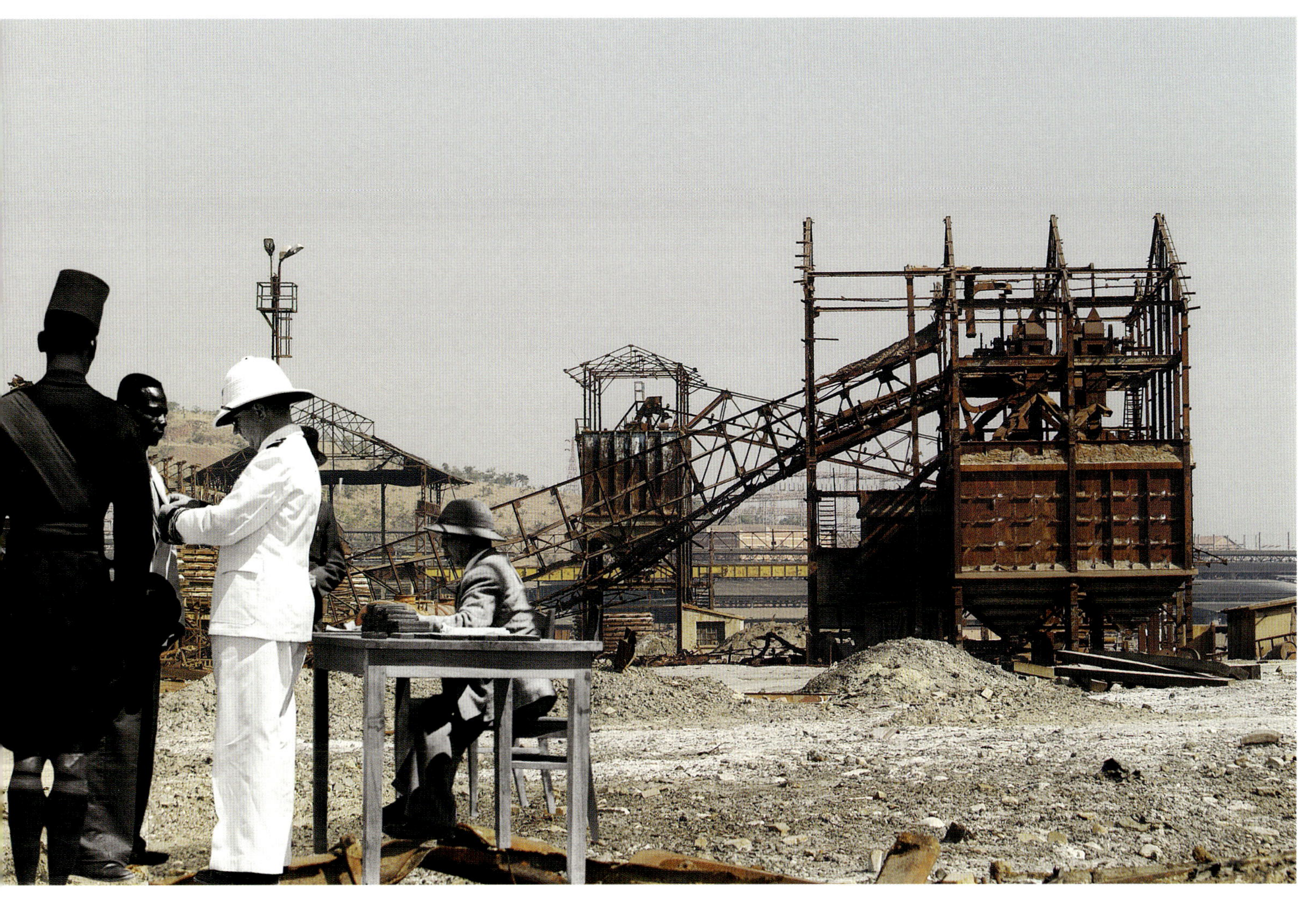

 Lisa Aronson

Rereading Landscape

Using photography as their medium, artists **Sammy Baloji** from the Democratic Republic of the Congo and **Lara Baladi** of Cairo, Egypt, are redefining and in other ways challenging their surrounding spaces. Baloji, who lives and works in the city of Lubumbashi, uses images to interrogate the cultural and industrial history of the Katanga region, one of the richest in the Congo because of its mineral resources. His most recent works, two of which appear in this show, revolve around a series of mines built and operated by the Belgian industrial complex, the Union Minière du Haut Katanga, active during the late colonial period, which became Gécamines (General Mines Society) a few years after Congolese independence. To Baloji, these now-abandoned structures linger as haunting ghosts of the colonial past. He emphasizes that perception all the more by photographing the empty mines in color and then superimposing over them cropped black-and-white images found in archival photographs dating to the colonial and early post-liberation periods, when the mines were thriving. These images included both European colonial administrators, sometimes collaborating with African leaders, and black laborers, some of them convicts. In juxtaposing these people over the now-abandoned mines, Baloji's photo collages offer sobering narratives on colonial history and its remnants in the present.

Lara Baladi also challenges viewers to broaden and rethink perspectives on the environment, in this case, Cairo, Egypt. The Cairo images that have captured most people's imaginations are of the pyramids. Baladi turns our attention instead to the tall, brick apartment buildings cropping up in great numbers along the city's contemporary outskirts. To avoid paying taxes, builders avoid completing these cookie-cutter, high-rise apartment buildings, yet people live in them nonetheless. At street level, the city's lack of infrastructure and services shows in the piles of garbage. Other photographs illustrate the developers' idealized billboards of completed buildings, which starkly contrast with the real unfinished ones. In her book titled *Hope*, in which all the images of her exhibition piece are included, Baladi writes, "The stark reality is that none of these towers of Babel is ever finished and that the green fields in which they

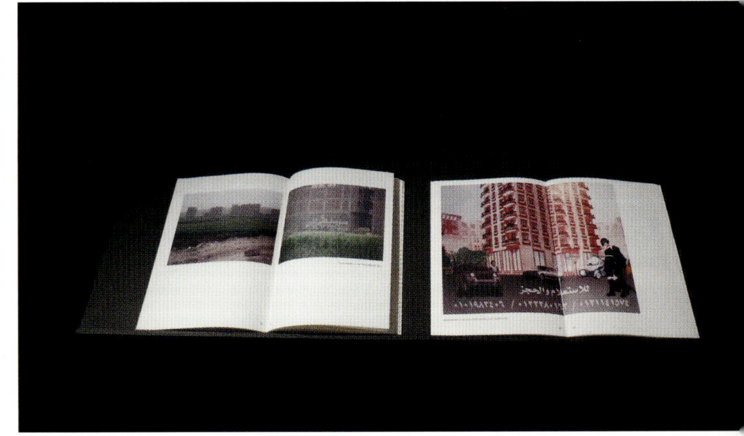

PAGES 56–59
LARA BALADI
Hope, 2010
Artist's book, photographs, and CD with *Donkey Symphony*
Book 7⅞ × 5⅞ inches

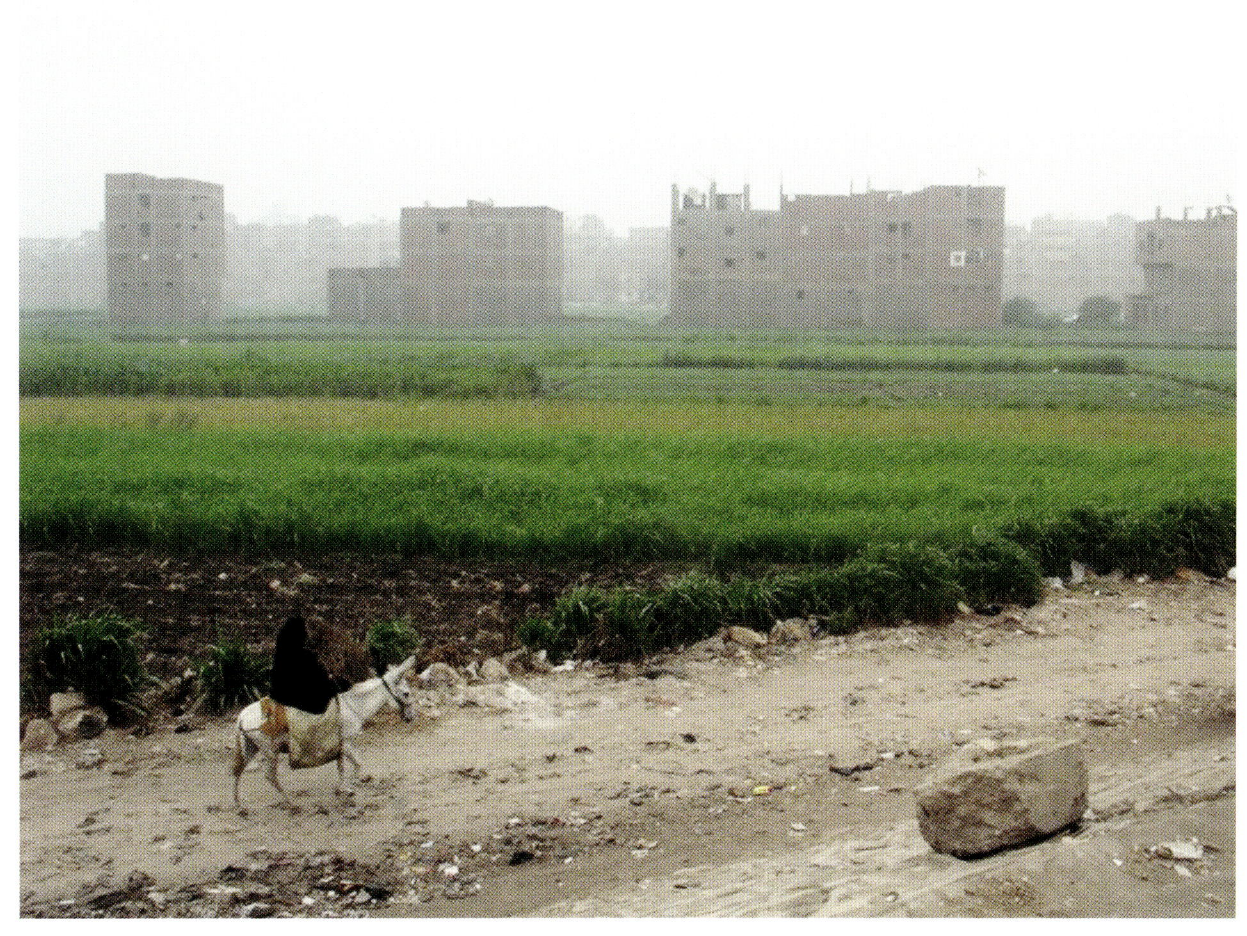

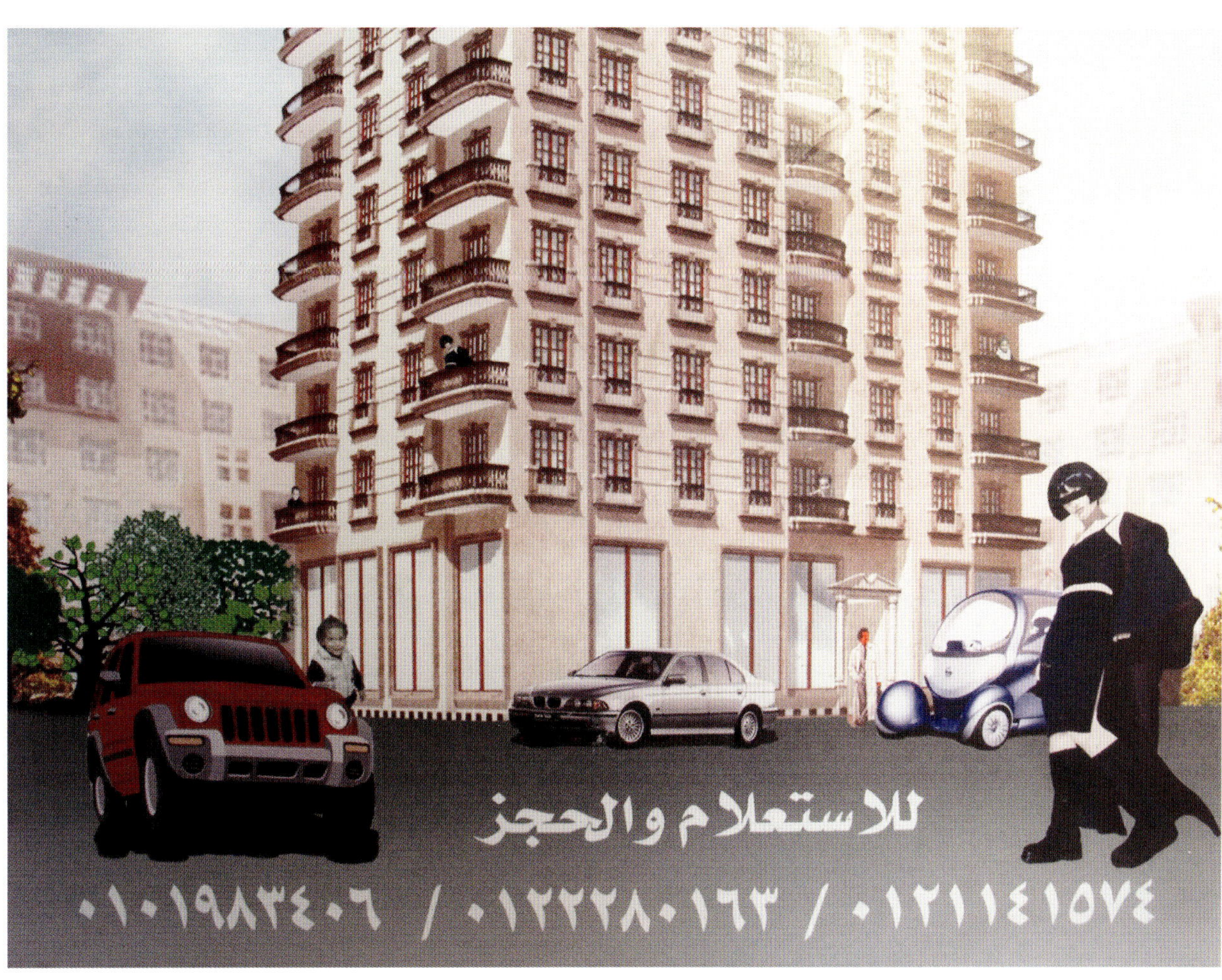

 Lisa Aronson

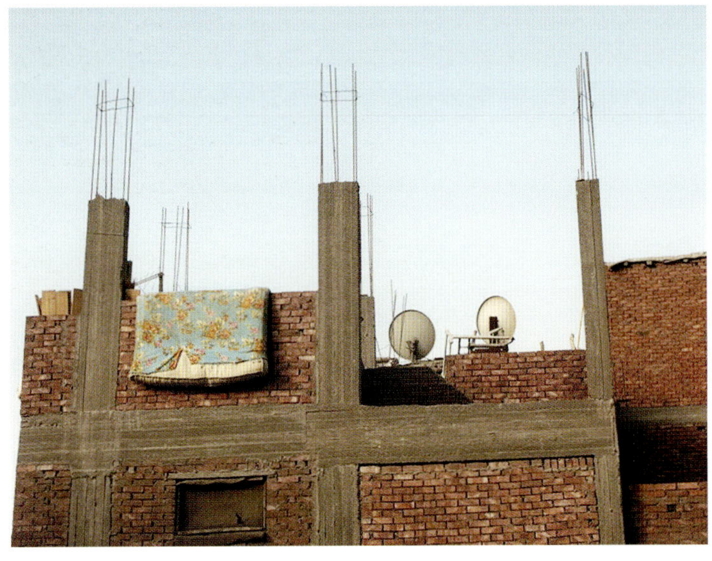
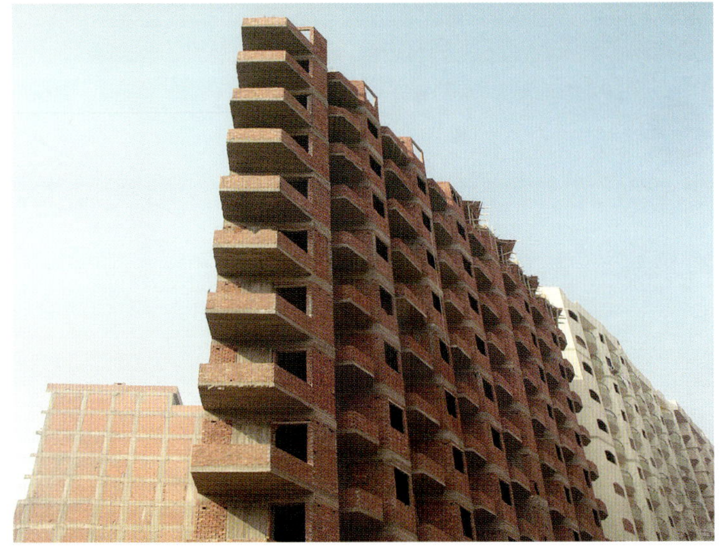
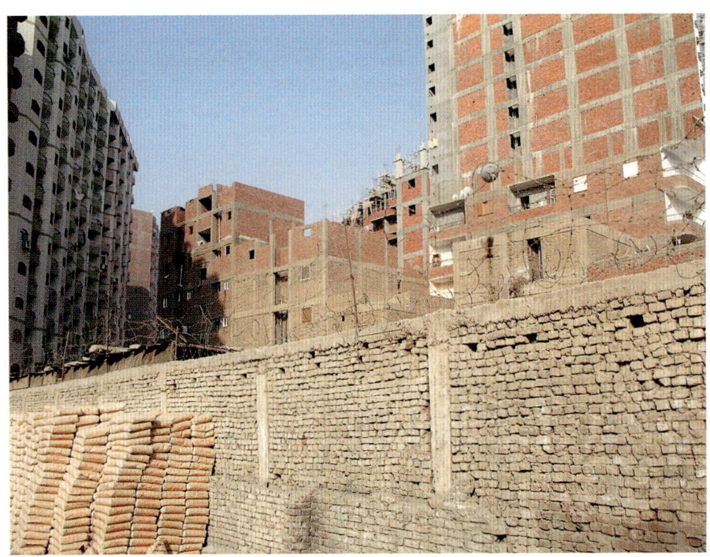
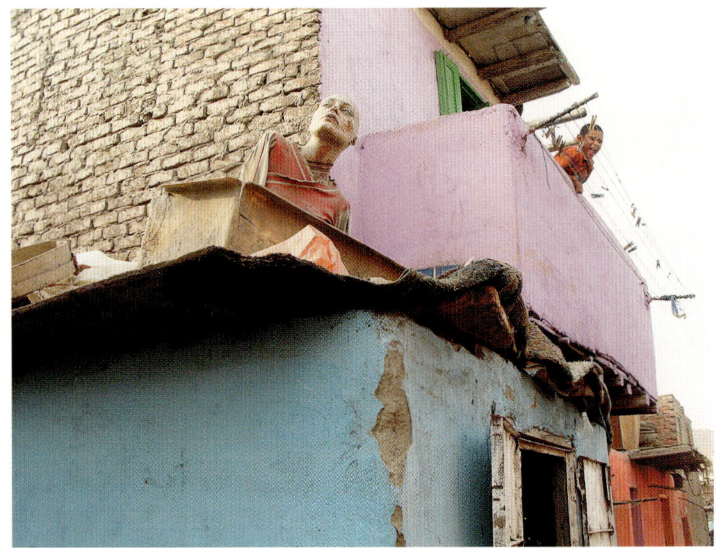
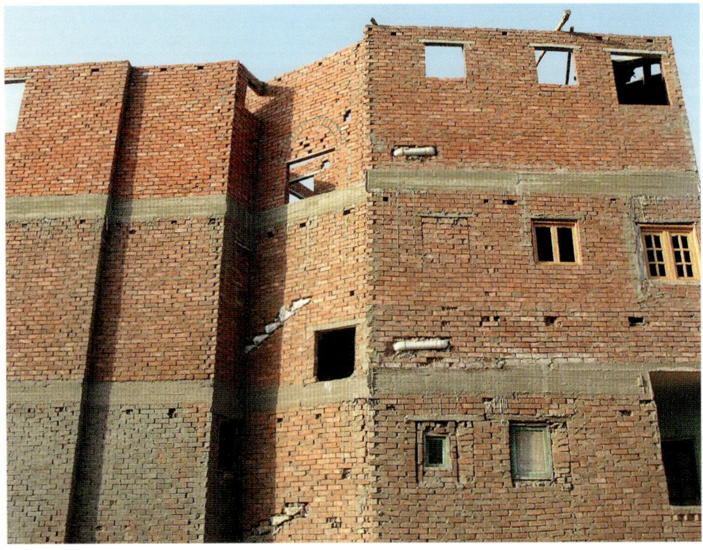
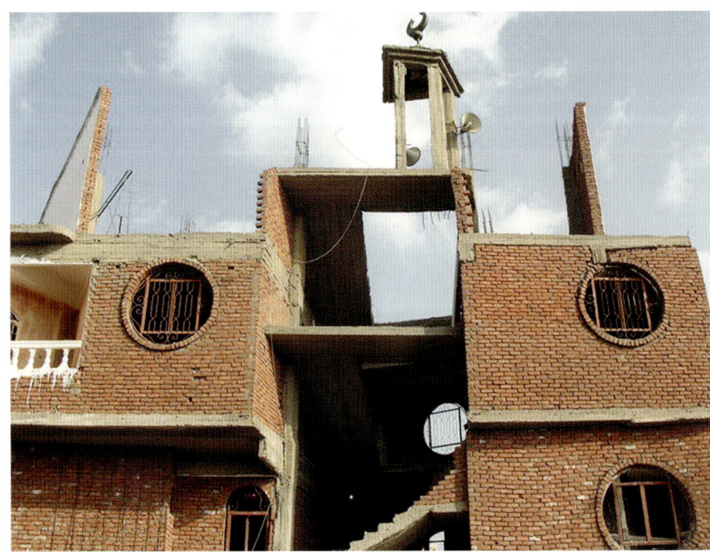

   Land, Landscape, and Contested Spaces

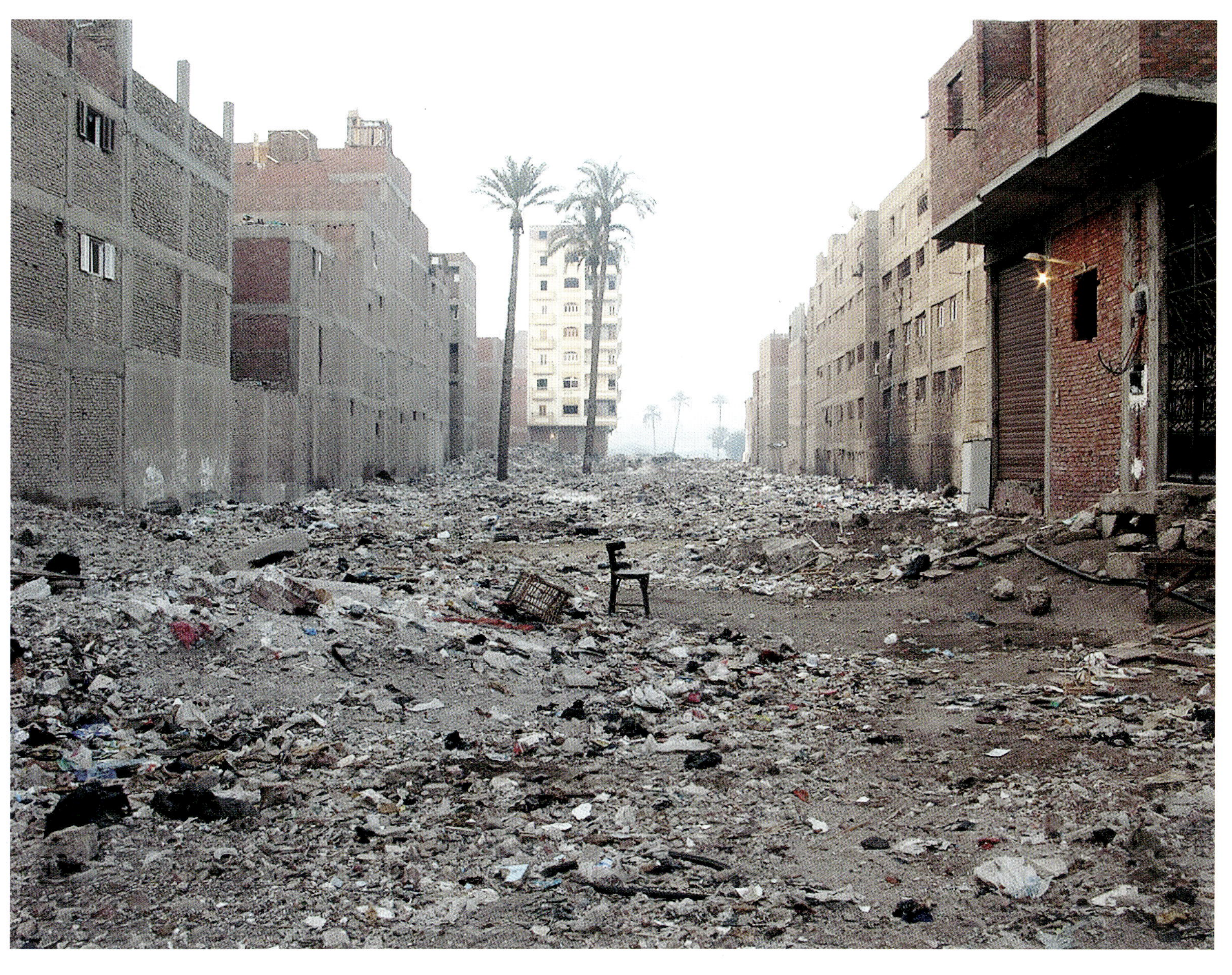

59   Lisa Aronson

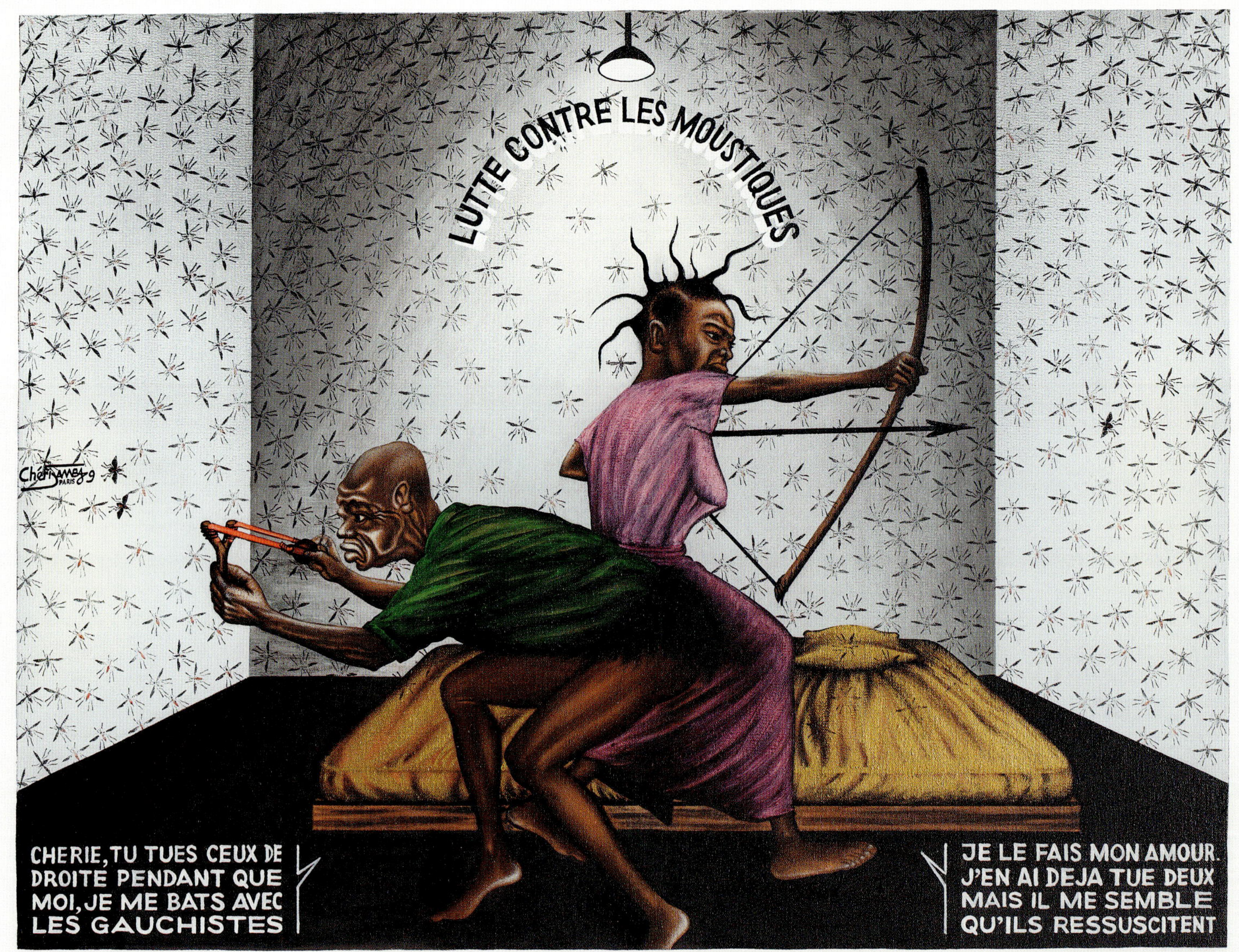

LUTTE CONTRE LES MOUSTIQUES
CHERIE, TU TUES CEUX DE DROITE PENDANT QUE MOI, JE ME BATS AVEC LES GAUCHISTES
JE LE FAIS MON AMOUR. J'EN AI DEJA TUE DEUX MAIS IL ME SEMBLE QU'ILS RESSUSCITENT
EN AFRIQUE, LA MALARIA TUE PLUS QUE LE SIDA. SURTOUT CHEZ LES PETITS ENFANTS. LE VIRUS DE LA MALARIA S'APPELE "MOUSTIQUES" CE VIRUS SEMBLE ETRE TRES PUISSANT QUE LES BLANCS ET LES NOIRS QUI VIVENT EN AFRIQUE

are planted are now urban sprawl and refuse."[15] The final photo in her display of images shows a single empty chair, upright in the middle of a soulless street strewn with trash. What fills Baladi's images, more than anything, is a coolly expressed but ultimately overwhelming anxiety for the future.

Landscape, Humor, and Irony

Almost all of Congolese **Chéri Samba**'s paintings involve social commentary; so it comes as no surprise that environmental issues engage him in his work. Samba is perhaps the most accomplished, and certainly most renowned, exponent of the "popular painting" tradition in the Democratic Republic of the Congo (DRC).[16] Through his often bitingly satirical paintings, Samba addresses almost every social ill imaginable, including AIDS (SIDA) and sexual promiscuity, political corruption, consumerism, and, as shown here, problems related to the environment. Drawing on his artistic background as a sign painter, Samba makes his painting as accessible as possible, combining imagery with legible, often humorous explanatory texts in French, English, and Lingala, the main language spoken in the DRC.

In his painting *Lutte contre les moustiques* [Fighting mosquitoes] (1989) Samba depicts the frustrating efforts a man and woman make to eradicate the ever-persistent mosquito, the primary carrier of malaria in Africa and a major threat to the lives of African people. The painted captions illuminate their frustration as they attempt to kill off the swarm of mosquitoes with a slingshot and a bow and arrow. Samba imagines an equally futile and far-fetched solution to the scarcity of water in his painting *Problème d'eau, ou trouver l'eau?* [The water problem, where to find water?] in which he addresses the scarcity of water by showing himself straddling a rocket to Mars with bucket in hand to seek a reliable water source. Again the absurdity evokes laughter, while the social critique remains serious.

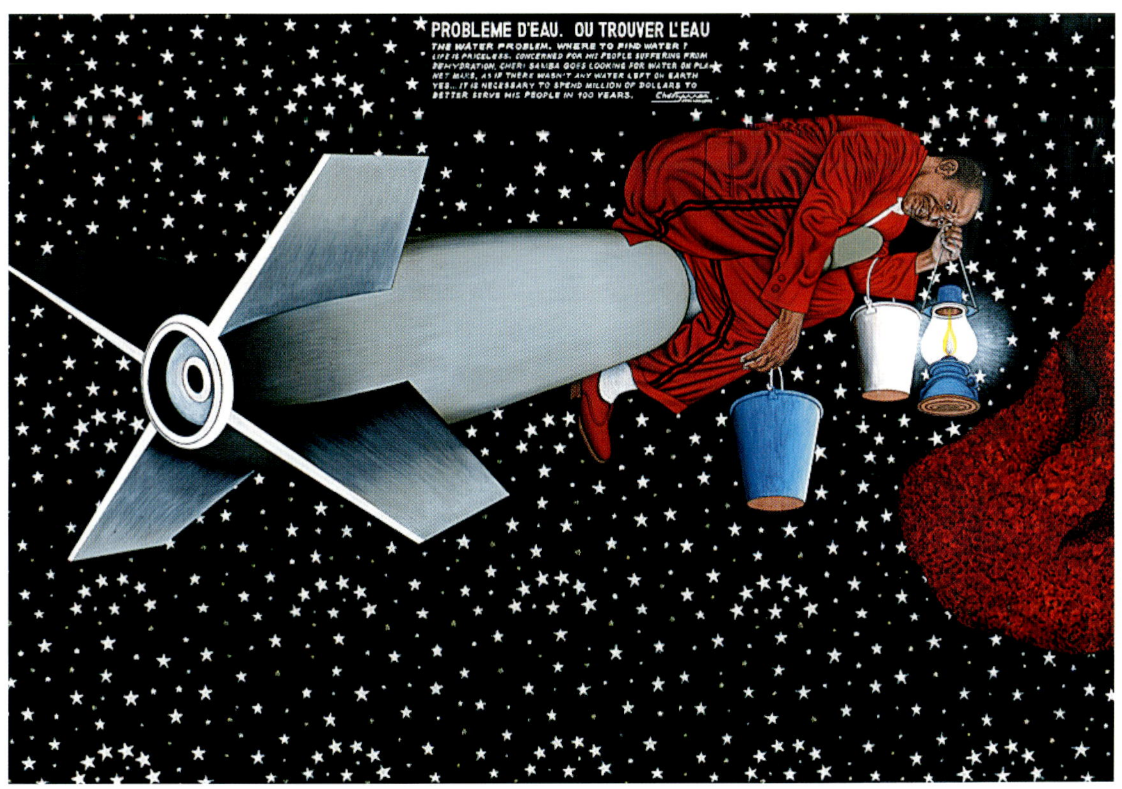

Barthélémy Toguo is as eclectic as artists come, with a repertoire of artworks including everything from painting, drawing, installation, performance and video, to photography, prints, ceramics, and collage. Regardless of medium, Toguo's art is concerned with human struggles within the larger global space, always with a degree of irony and humor that evokes African modes of expression, from traditional masquerades to public protest.[17] Humanity's relationship to nature runs throughout his work as a leitmotif. In his watercolors, he shows flowers and other vegetation growing out of human bodies. In one case, the flowering body itself lies half buried in the ground, as though sprouting from the earth.

Concerns about the land also come through in his *Projet Agricole* [Agriculture project], which is embedded in his larger *Bandjoun Station Project*. The latter involved the construction of a cultural center in Bandjoun, financed with Toguo's own money and designed to serve a variety of functions: museum, art studio, theater, and seminar space for public lectures. For the agricultural component, as he writes on his website:

In order to expand the parameters of this ambitious artistic and cultural building protect, I decided in addition to use three hectares of land for a project that integrates artistic and agricultural elements. This way of combining both environmental issues and social experimentation sets an example for the local youth; it is a way of creating a dynamic and equitable relationship between guests and hosts, as well as of demonstrating the central importance of agriculture in our quest to become self-sufficient in food production.[18]

The agricultural component would provide the supportive environment he believes young artists need in order to learn about art and to create.

At least two of the three Toguo photographs exhibited in this show engage the idea of landscape, either referentially, or in a conceptual way. His photograph *Stupid African President 3 (Afrika Oil)*[19] is connected to his performance at the State Museum of Contemporary Art in Thessaloniki, Greece, in 2007, during which he crouched in a barrel marked "Oil" while guzzling the contents of a plastic bottle within a matter

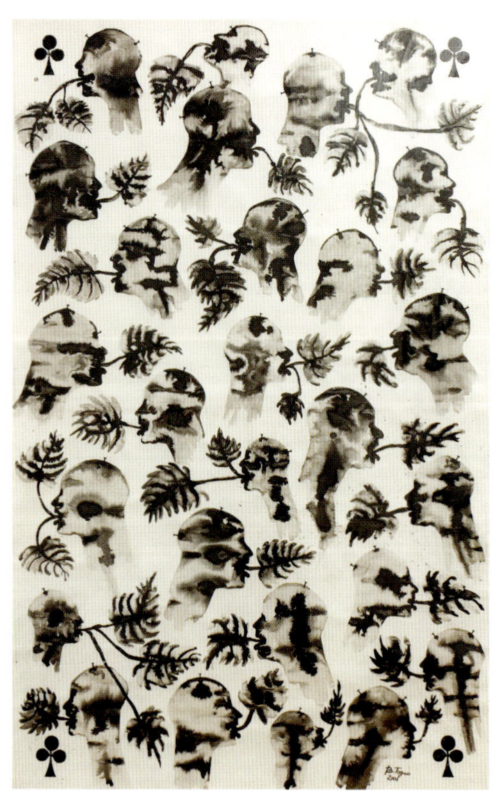

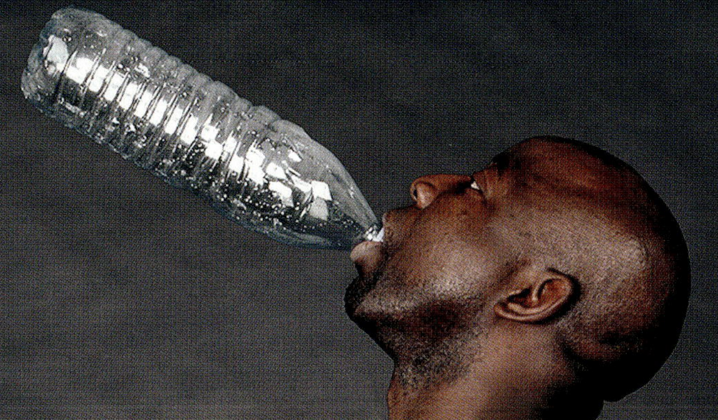
AFRIKA
OIL ?

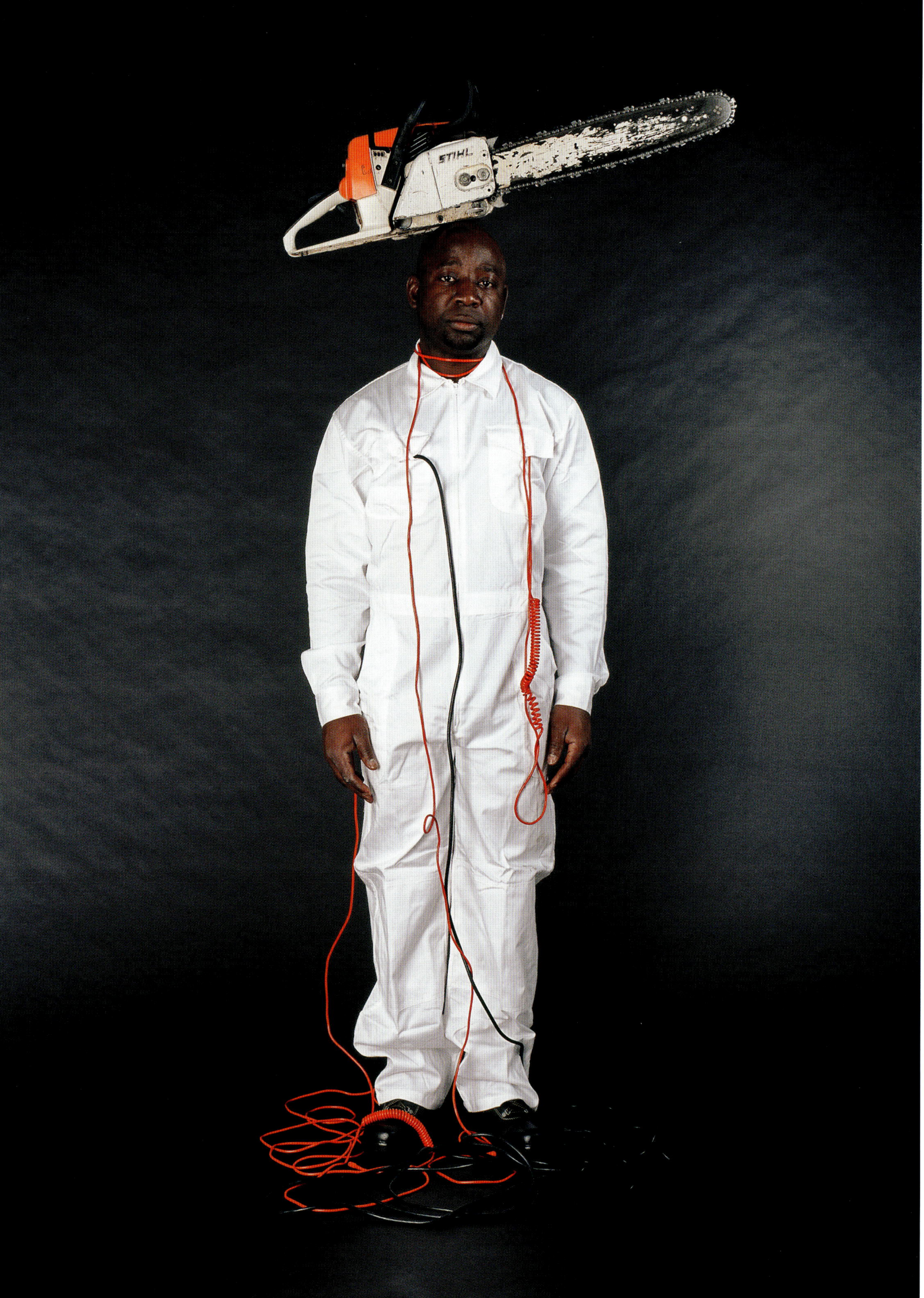

of minutes. The title of the performance, *The Bitter Destiny*,[20] hints at his environmental concerns, including the economy's unhealthy but unquenchable thirst for oil.

The accomplices Toguo sees as responsible for destroying the African landscape include African presidents, the focus of the *Stupid African President* series. *Stupid African President 1* shows a smartly attired Toguo standing before a map of colonial Africa, as if addressing some topic of serious continental relevance. The possibility of taking this

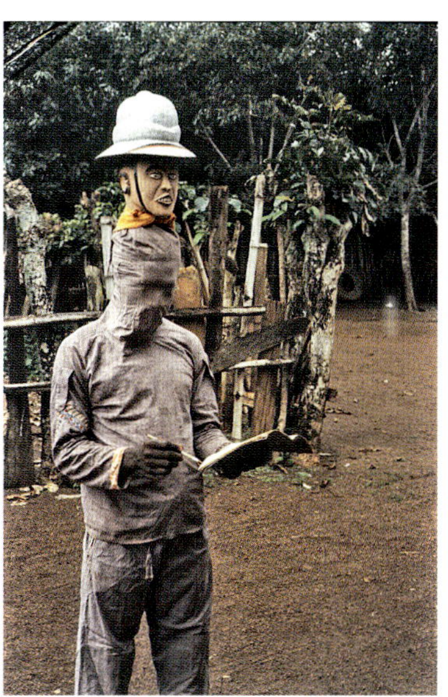

image as a straightforward representation of African leadership is undermined not only by the title, but by its pendant image, *Stupid African President 2*, in which the artist wears a rumpled white coverall and balances a chain saw ludicrously on his head. The image evokes laughter similar to the way Igbo maskers would when wearing a mask of the colonial officer as he takes copious notes on the people he is colonizing. Though comically presented, Toguo's environmental concern is evident, for the portable chain saw has enabled widespread destruction of the African forest.

Conclusion

In the introduction to their edited volume *Social History and African Environments*, William Beinart and Joann McGregor argue that "the integration of scientific, historical and social science approaches is essential for an understanding of environmental change, and for strategies of regulation."[21] This essay adds to their list the voices of African artists, whose powerful and often provocative works deploy everything from biting satire to lyrical beauty to offer fresh, new perspectives on prevailing political, economic, and social implications of ecological destruction and exploitation. Though these artists come from diverse spaces and individual histories within the African landscape, they share a concern for the plight of the African environment, asking humanity to take notice of its problems, and work to overcome them.

     Lisa Aronson

1. Jerry Buhari in correspondence with Michelle Lorenzutti, Skidmore College '09, April 23, 2009.

2. Buhari notes that British and American faculty encouraged and taught this landscape-focused curriculum.

3. Author's communication with Jerry Buhari, July 27, 2010.

4. For more about Jerry Buhari's work, refer to his website: www.jerrybuhari.com (accessed August 7, 2011).

5. Buhari's accompanying poem by the same name offers a poignant literary account of the human consequences of competition over oil:

IN THE PIPELINE

By early morning
The news had gone out
There is a broken pipeline
Somewhere near the village
Within a few hours excitement has
    swept the entire village
Men, women, the youth of the village
Even the aged and children
Ran down to get their share of
    what was
In the pipeline.

Suddenly there was an explosion
Screams, then silence … and smoke
What remained are scattered mangled
    charred bodies
Or what looked like bodies
The dividends of demo*crazy* often
    stay too long
In the pipeline.

"In the Pipeline," his title for both the poem and the painting, has broader political implications for citizens of Nigeria, whose presidents often invoke this phrase when offering seldom-met promises of reform.

6. Jeremy Foster, *Washed with Sun: Landscape and the Making of White South Africa* (Pittsburgh: University of Pittsburgh Press, 2008), 80.

7. See Rooke Gallery website for commentary on his work in his show *Study of Trees*, Rooke Gallery, Newtown, Johannesburg, October 28–December 17, 2010. http://www.rookegallery.com (accessed August 7, 2011).

8. See, for example, the American Museum of Natural History's website detailing James P. Chapin and Herbert Lang's 1909–1915 Congo Expedition, http://diglib1.amnh.org (accessed August 7, 2011).

9. Isolde Brielmaier, "A Conversation with Zwelethu Mthethwa," in *Zwelethu Mthethwa* (New York: Aperture, 2010), 93–94.

10. Kim Miller, "Moms with Guns: Women's Political Agency in Anti-Apartheid Visual Culture," *African Arts* 42/2 (Summer 2009): 68–75.

11. Liz Hoggard, "Sweeping," in *Sweeping: Sculpture by Sokari Douglas Camp* (London: Douglas Camp Publications, 2006), 3.

12. Ibid, 3.

13. Ogoni poet, author, and environmental activist Ken Saro-Wiwa (1941–1995) was an outspoken critic of the Nigerian government, which he faulted for not enforcing regulations to prevent the destruction of his home environment in the Niger Delta. Although Saro-Wiwa's approach was consciously non-violent, the Nigerian government arrested and tried him, and in 1995 hanged him with eight other activists. The consequent inter-national outrage resulted in Nigeria's suspension from the Commonwealth of Nations for more than three years.

14. In 2005, PLATFORM, an organization that uses the arts to highlight oil-industry abuses worldwide, joined with Amnesty International, the Arts Council of Britain, and Greenpeace to launch a design competition for a Saro-Wiwa memorial to mark the tenth anniversary of his execution. Douglas Camp's memorial sculpture won this competition.

15. Lara Baladi, *Hope* (Cologne, Germany: Rautenstrauch-Joest-Museum Kulturen der Welt; Saratoga Springs, NY: the Frances Young Tang Teaching Museum and Art Gallery at Skidmore College, 2010), 5.

16. Bogumil Jewsiewicki coined the term *popular painting* to describe the type of painting produced by artists from urban areas of Zaire (now the Democratic Republic of Congo) in the 1970s. Urban artists, such as Chéri Samba, typically combined a social realist style of painting with text to address themes of social injustice and chronicle social and political memory. Bogumil Jewsiewicki, "Painting in Zaire: From the Invention of the West to the Representation of Social Self," in *Africa Explores: Twentieth Century African Art*, ed. Susan Vogel (New York: Center for African Art, 1991), 130–151.

17. Mary Jo Arnoldi, "Masks and Masquerades," in *African Folklore: An Encyclopedia*, ed. Philip M. Peek and Kwesi Yankah (New York: Routledge, 2004), 244; Achille Mbembé, "Provisional Notes on the Postcolony," *Africa: Journal of the International African Institute* 62/1 (1992): 6.

18. Toguo goes on to say, "It is a strong political statement; our artistic breeding ground will in turn give birth to a coffee plantation, a critical act which magnifies the artistic gesture and powerfully critiques what Léopold Sédar Senghor has called 'the deterioration of the terms of exchange, in other words the lowering of prices for export imposed by the West which punishes and impoverishes the farmers of the South.'" http://www.barthelemytoguo.com/3projets/bandjoun_station/bandjoun.html (accessed August 7, 2011).

19. On the artist's website, this work has been listed as simply *Afrika Oil*, but according to Toguo it is part of the *Stupid African President Series* and was published as such in *The lost dog's orchestra* by Galerie Lelong (Paris, France, 2010).

20. http://www.barthelemytoguo.com/20euvres/3performances/africanoil.html (accessed August 7, 2011).

21. William Beinart & Joann McGregor, eds., *Social History and African Environments* (Oxford: James Currey; Athens: Ohio University Press; Cape Town: David Philip, 2003), 2–3.

Karen Kellogg

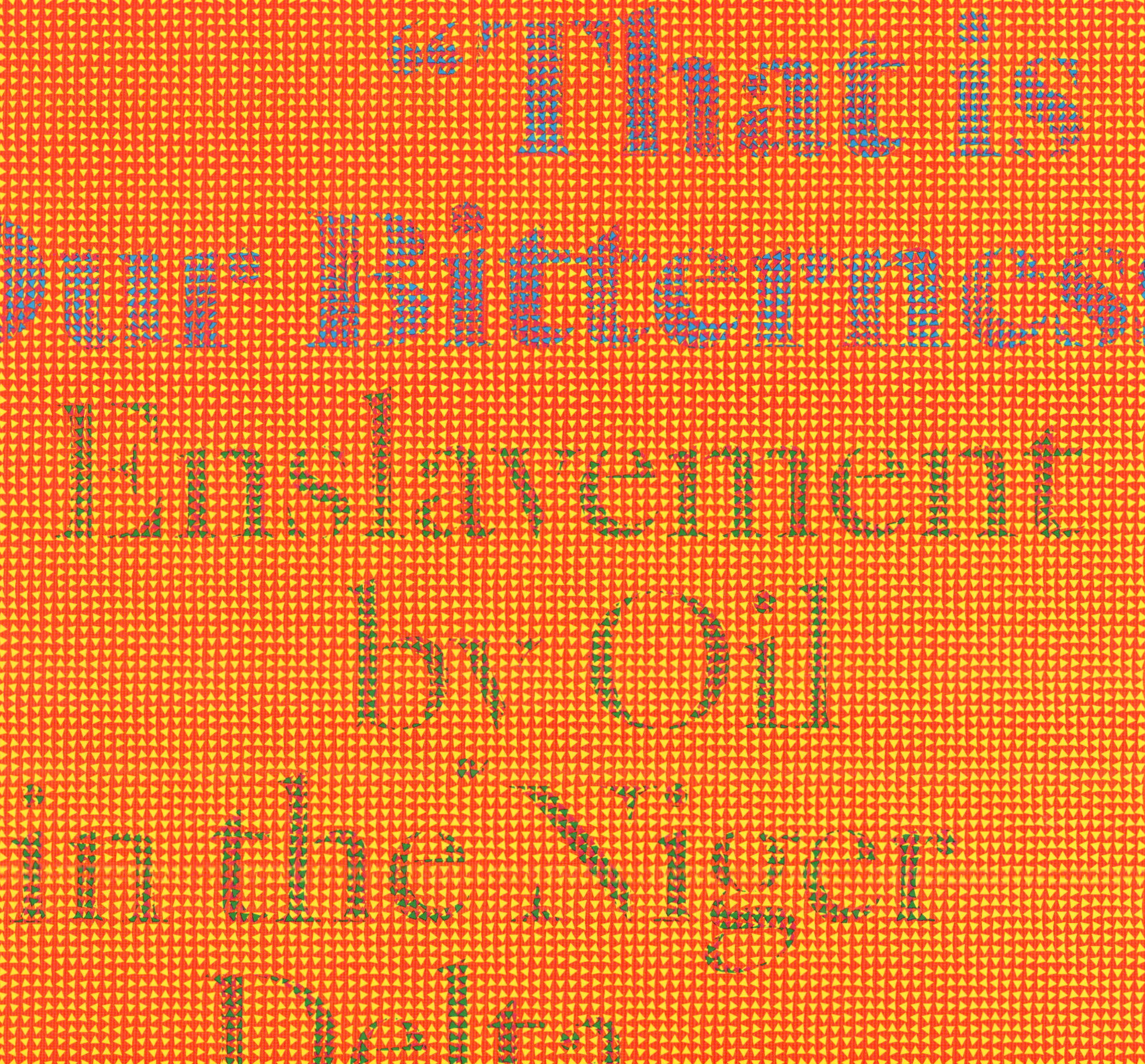

"That is Our Bitterness"

Enslavement by Oil in the Niger Delta

global economic growth from the seventeenth through the nineteenth centuries is a remarkable human achievement, with spectacular spurts accentuating a general escalation of per-capita world gross domestic product (GDP), from $565 in 1500 to $5,709 in 1998 (in 1990 international dollars).[1] With the exponential growth in GDP came, of course, unprecedented prosperity, and, according to global indicators, we now live longer, healthier, more literate lives. But as any statistician will tell you, averages are only one piece of any analytical story, and the more noteworthy features often lie in the variation around the average. In many ways, *Environment and Object · Recent African Art* explores a deep and complex valley in the topography of global wealth and examines the backs that carried an unjustly heavy load in building that economic growth.

A foundation of all economic development is energy, and while sporadic energy sources afford a certain amount of growth, stored energy provides the continuity necessary for the type of economic advancement that has defined our modern world. Before the Industrial Revolution, we primarily had at our disposal the muscle power of the human body and domesticated animals, and societies advanced their wealth using a somatic energy regime. Since the human body is more efficient than draft animals at converting chemical to mechanical energy, stockpiles of humans became a proficient means of "making the rich and powerful, richer and more powerful."[2] While slavery acted as the flywheel to society's energy storage through the eighteenth century, the Industrial Revolution inaugurated a very different energy system based primarily on fossil fuels. Our world energy use skyrocketed from 250 million metric tons of oil equivalents in 1800 to 800 million in 1900, with an even more gigantic leap to 10,000 million by 2000,[3] and fossil fuel use began to correlate tightly with gross national product (GNP) (Figure 1). By 1950, any society that did not consume vast quantities of fossil fuels was

Figure 2
Top sub-Saharan
African oil production
by country,
December 2006

● 10,000 barrels
per day

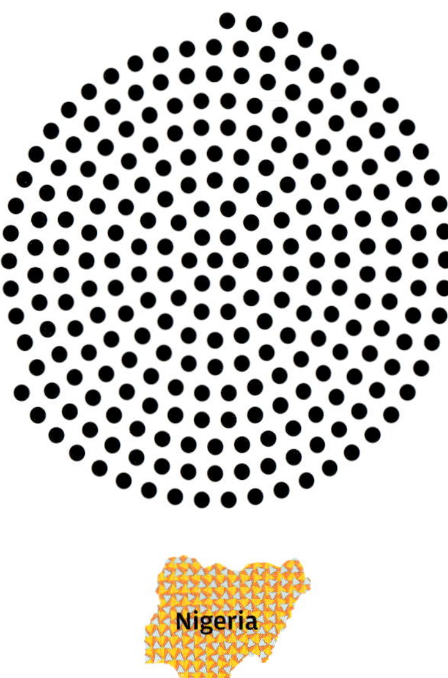

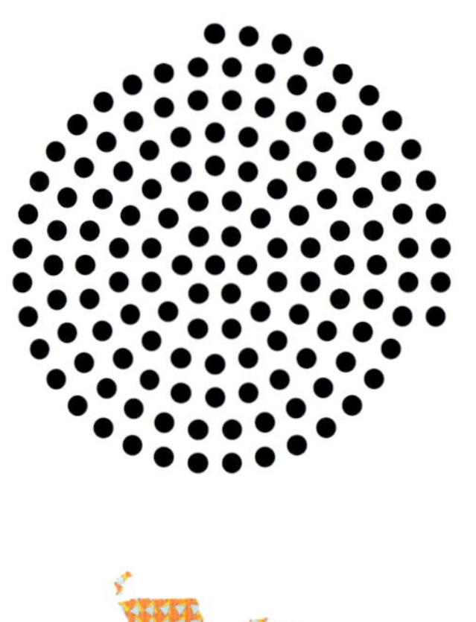

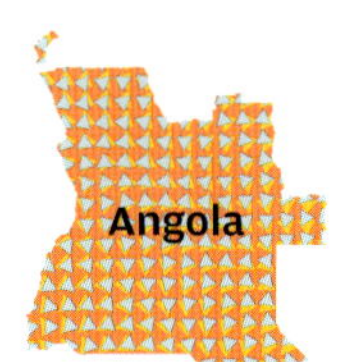

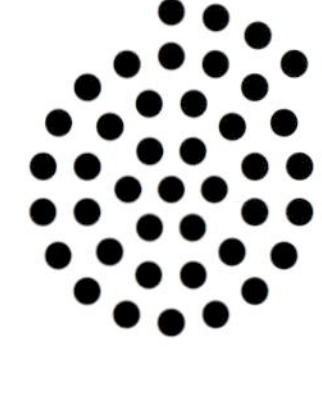

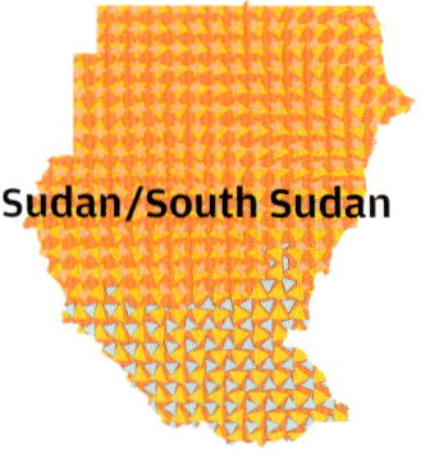

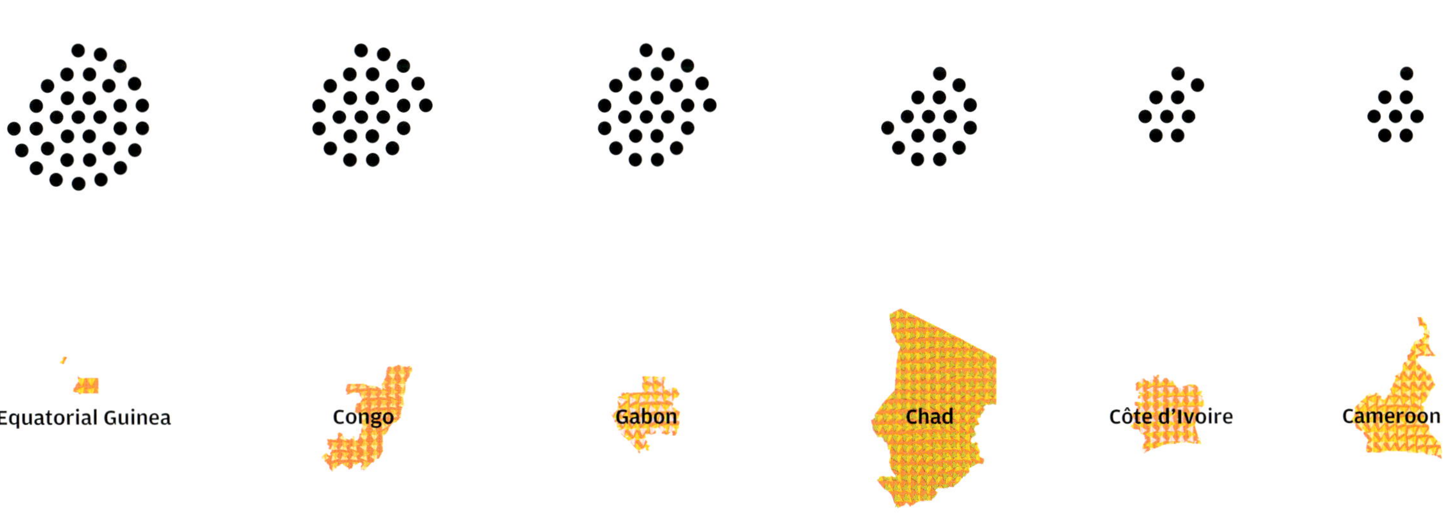

Karen Kellogg

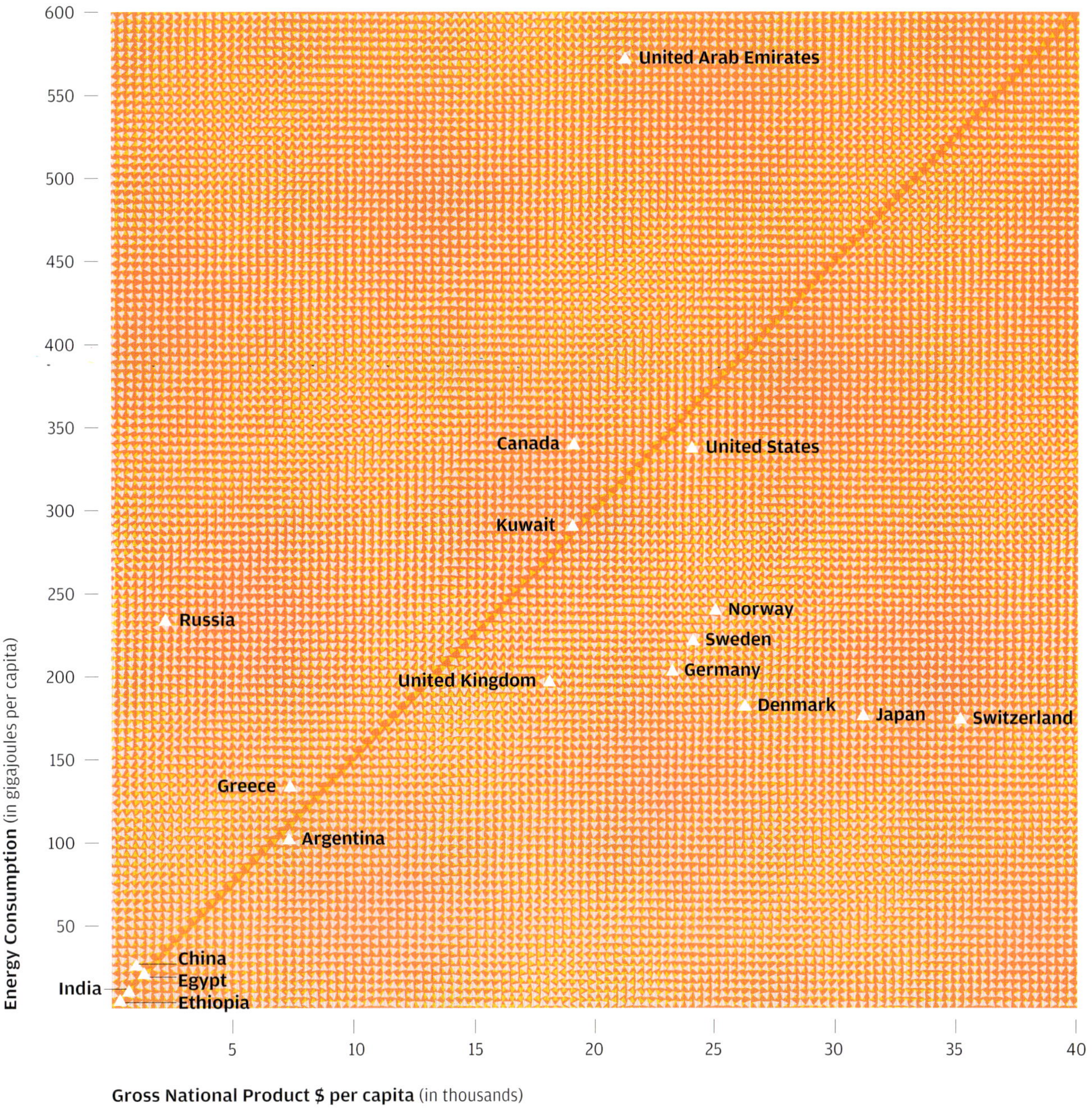

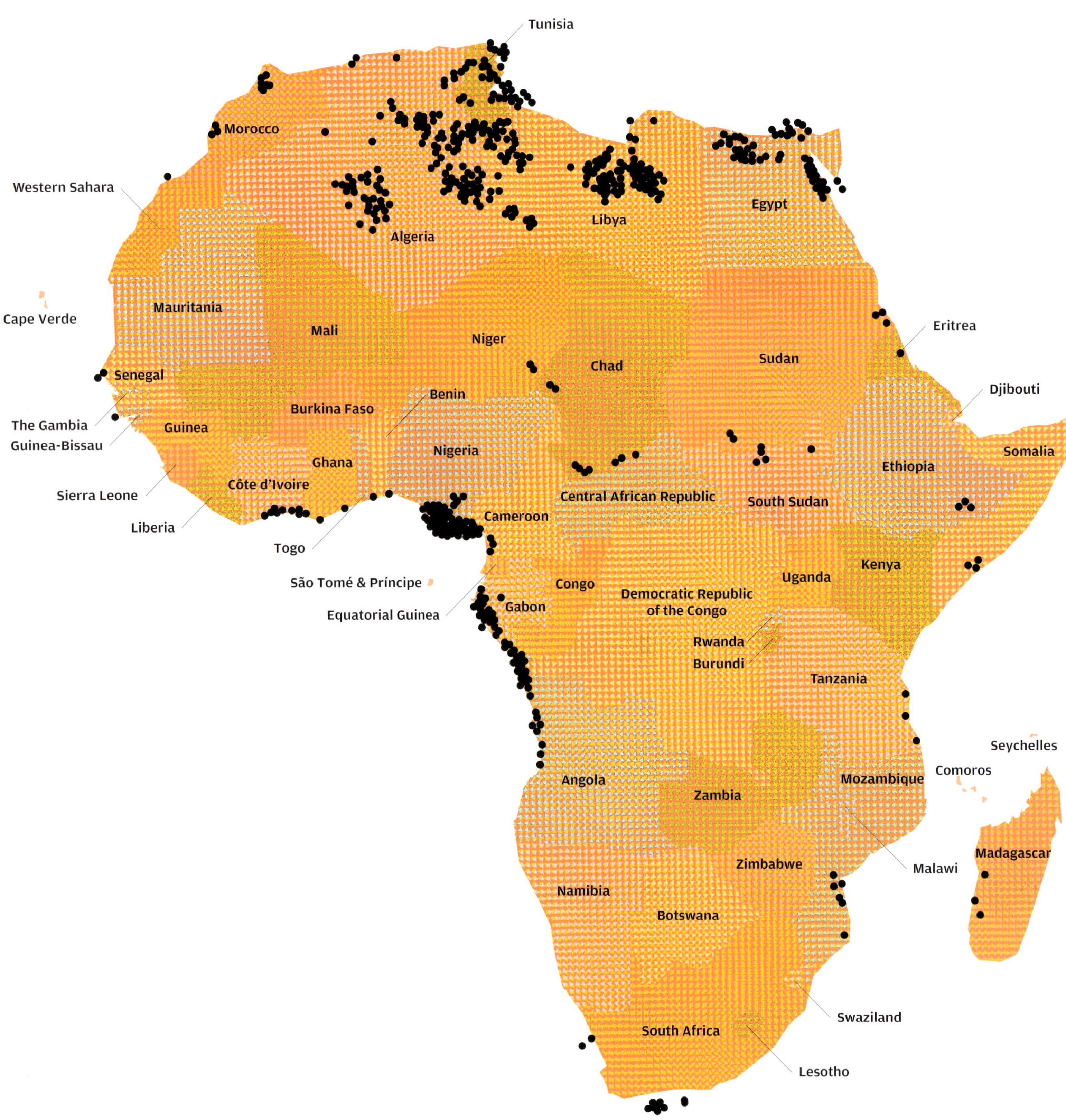

"That Is Our Bitterness"

condemned to poverty.[4] While coal was the primary fossil fuel of the 1800s, oil quickly became the fuel of choice in the 1900s because of its relative ease of transport and use.[5] Oil remains the mother of all fossil fuels—and commodities for that matter—and our global economies still rest firmly on the belief that the vast pipelines of cheap oil will continue to flow.

Both the somatic and fossil-fuel regimes had deep roots in Africa, and Nigeria, in particular, shouldered a disproportionate burden of both. While in 1500 African people were a minority of the world's slave population, by the end of the seventeenth century, they had become the majority. As many as twelve million of the eighteen million slaves taken from Africa in the seventeenth through nineteenth centuries came from the western part of the continent, with Nigeria among the hardest hit.[6] Similarly, Nigeria ultimately became a major exporter of oil, in fact, the largest in sub-Saharan Africa (Figure 2). Although many large oil fields were discovered in Africa, only the reserves of northern Africa played a major role before 1973, with major exports from both Algeria and Libya beginning in the mid-1900s (Figure 3). In 1973, the Organization of Petroleum Exporting Countries (OPEC) restricted oil production, inadvertently launching another push for oil exploration and extraction, and, as a result, sub-Saharan Africa became a major player in the production of black gold (Figure 3). Nigeria was in an especially strong position to ramp up oil production after the OPEC restriction, largely owing to strategic efforts by Dutch Shell and British Petroleum (Shell-BP); in fact, the 2.4 million barrels per day of Nigerian crude-oil output in 1973 is comparable to present-day production.[7]

Nigeria's oil lay in the Niger Delta, and Shell-BP, which had been granted licenses by the British colonial government to explore the delta, struck oil in 1956. In the feverish decade that followed, Shell-BP tactically constructed extensive infrastructure to extract and process the highly desirable light, or "sweet," crude of the delta, and prudently backed the triumphant central government in the civil war of 1967–1970.[8] To date, over 6000 wells have been sunk to tap oil in the Niger Delta, and 606 oilfields, 7000 km of pipelines, 275 flow stations, 15 export terminals, and 4 refineries now largely define the physical landscape (Figure 4).[9] Oil has greased the political economy of Nigeria, and the oil sector now accounts for approximately 95 percent of foreign exchange earnings and 80 percent of budgetary revenues.[10] Nigeria has climbed to the eleventh largest producer and the eighth largest exporter of crude oil in the world,

          Karen Kellogg

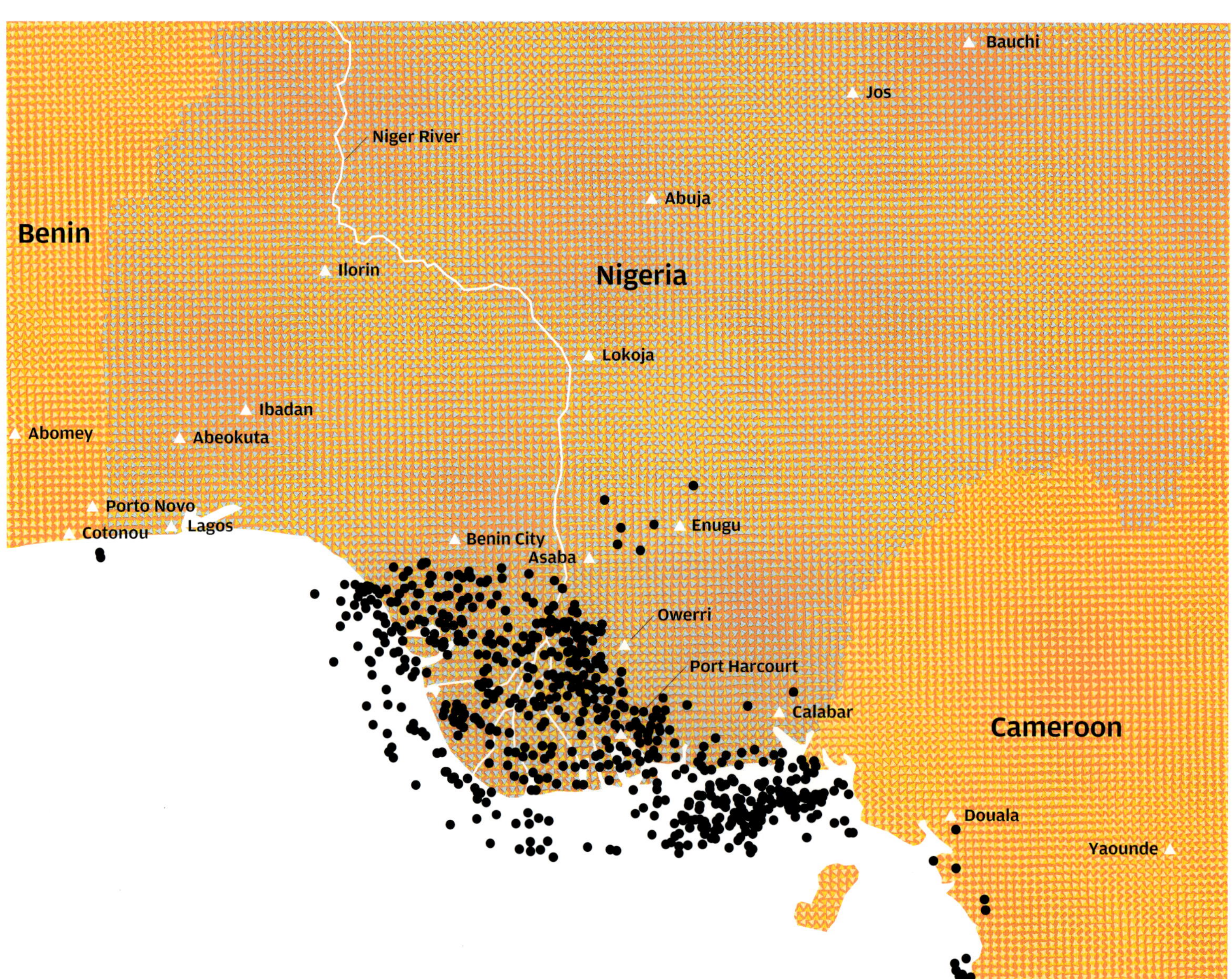

Figure 4
Oil and gas fields in the
Niger Delta, 1996

and sizable proven reserves guarantee that oil will be a strategic interest to Nigeria and the global community for years to come (Figure 5).[11] The United States, incidentally, relies on the Niger Delta region for approximately 10 percent of its oil imports,[12] which are predicted to increase to 20 to 25 percent by 2020.[13]

So what has Nigeria gained from participating so heavily in two critical global markets that have fueled the economic advancement of so much of the world? Some Nigerians occupied lucrative, pivotal roles in the slave trade, but the costs for most Nigerians were deep beyond words; they literally paid with their lives. The chronicles of the oil economy are actually hauntingly similar to those of the slave trade, exacting long-term economic, environmental, and social costs that continue to enslave many Nigerians. While, again, a small number of Nigerians enjoy the wealth that flows from oil exportation, the majority suffer in extreme poverty. Although the Nigerian exchequer takes in an estimated $1.5 billion in oil revenue every week,[14] Nigeria continues to rank below countries like Haiti on the Human Development Index, a synthesis metric of life expectancy, income, and educational attainment.[15] In fact, instead of fueling broad prosperity in Nigeria, the growth in oil production is associated with a decrease in per-capita annual income from $250 to $212 between 1965 and 2004 and with an increase in extreme poverty: the number of people existing on less than one dollar per day grew from 19 to 90 million (36 percent and 70 percent of the population, respectively).[16]

Such bewildering negative correlations are actually somewhat typical of the exploit-and-abandon mentality of extraction economies. Unlike manufacturing-based economies, extraction-based economies require smaller workforces and less complex

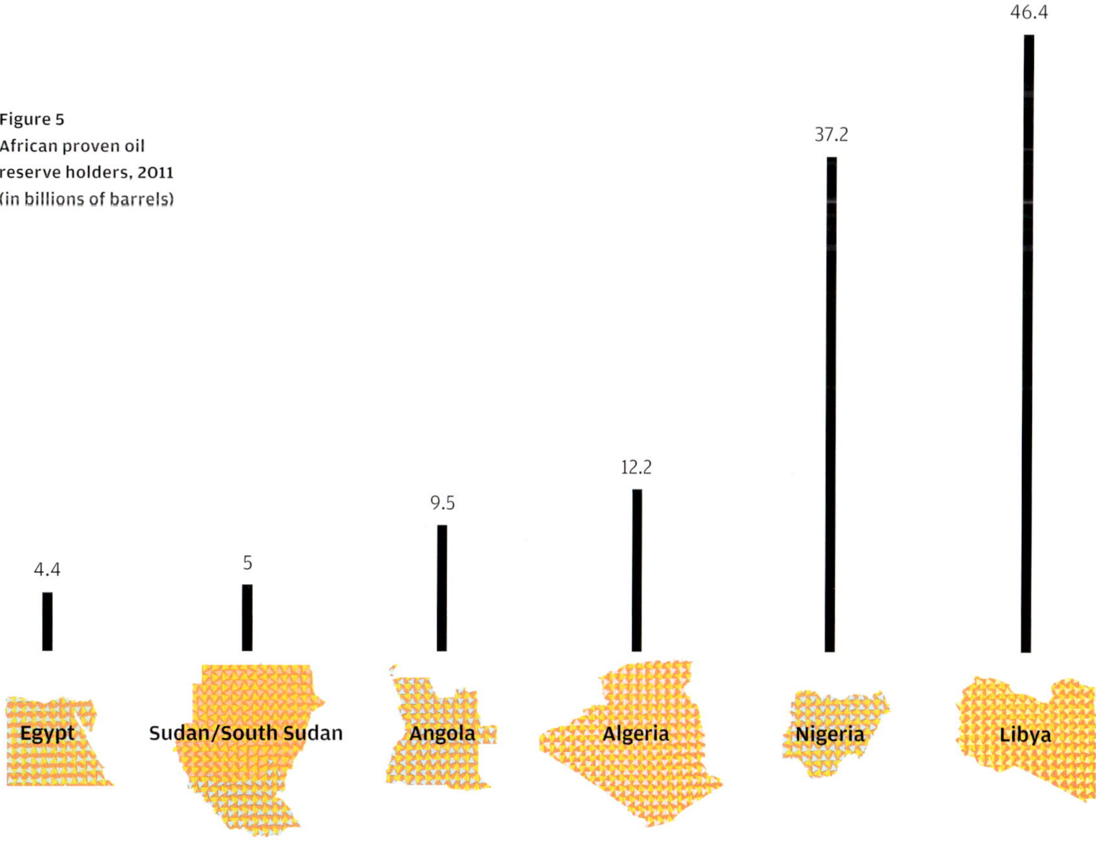

Figure 5
African proven oil
reserve holders, 2011
(in billions of barrels)

          Karen Kellogg

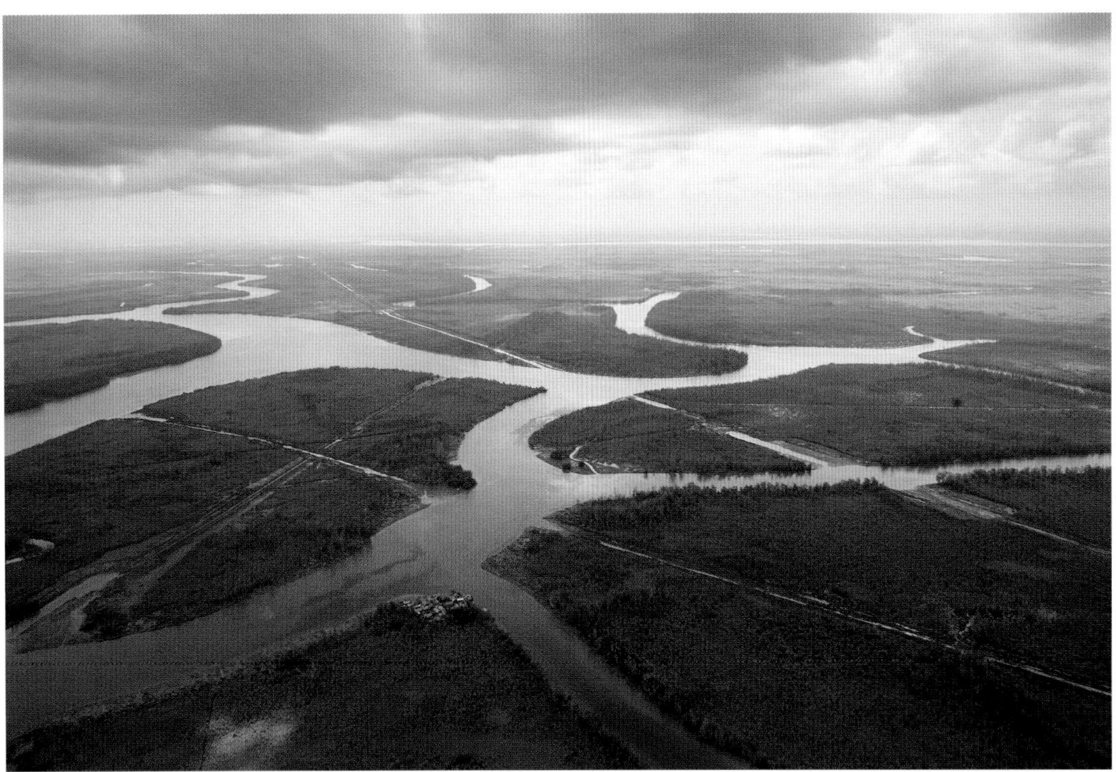

divisions of labor and hence foster little investment in education.[17] In fact, an educated
workforce might actually be viewed as a liability. Perhaps not surprisingly, then,
only 68 percent of the Nigerian population is literate, school life expectancy is only eight
years, and education expenditures make up a mere 0.9 percent of the GDP (as a point of
comparison, education expenditures in the United States are 5.3 percent of our much
larger GDP).[18] In addition, the large capital investments, the complex and volatile pricing
algorithms, and the domination of large multinational corporations make entry
into the oil industry nearly impossible, and, predictably, power and wealth have become
concentrated in the hands of a few.

The inequities associated with extraction-based economies often become even
more exaggerated by corruption, especially in environments where governments control
the rents.[19] A string of governments paved the way for Nigeria to become the poster
child for both extreme corruption and hyper-inequity. Oil royalties splashed through
state coffers, with at least $100 billion of the $600 billion in oil revenues accrued
since 1960 simply missing, and Nigeria was repeatedly identified as having one of the

     "That Is Our Bitterness"

most corrupt governments in the world by Transparency International, a nonprofit organization that develops and administers quantitative diagnostic tools of transparency and corruption.[20] As a result, approximately 85 percent of the oil revenues rest in the pockets of a mere 1 percent of the population.[21]

Once established, inequities run intergenerationally, often gaining momentum through economic entrenchment, with the less fortunate having access to fewer and fewer opportunities to change their circumstances and the more fortunate clinging to their trajectory for fear of having to live like the other side. With large-scale hydrocarbon industries also come significant environmental opportunity costs that reinforce a burgeoning economic divide. In the case of the Niger Delta, extensive air, water, and soil-water pollution make other forms of economic development challenging, considering that all long-term economies are founded, to a certain extent, on healthy ecosystems.

The biological wealth of the Niger Delta was, at one point, quite vast. Complex hydrologies fed extensive rainforests, brackish swamps, and mangrove forests, all filled with a rich array of plants, insects, reptiles, amphibians, birds, and mammals (Figure 6).[22] In very practical terms, biodiversity in any environment helps fulfill basic human needs by providing food, shelter, clothing, and medicines and also gives rise to various essential ecosystem services, including air and water purification, climate regulation, waste decomposition, and flood regulation. Biodiversity also bequeaths deeper meaning to human lives by providing sources of cultural, spiritual, recreational, linguistic, and aesthetic inspiration.[23] Intact, functioning ecosystems are the natural capital upon which all human-made capital depends, and it is now a widely held view that durable industries and employment opportunities are contingent on a reevaluation of current trends of environmental destruction.[24] While numerous activities, including agriculture and overpopulation, have contributed to ecological degradation in the Niger Delta, oil activities have borrowed significantly against the region's natural capital.[25] The delta in Nigeria is now considered the most ecologically endangered region in the world.[26]

The damage from oil operations in the Niger Delta is both chronic and cumulative, although assessing the full footprint of the oil industry is challenging because environmental-impact assessments are not often done or made fully public and because

     Karen Kellogg

scientific studies have been relatively few and far between. In broad strokes, however, oil exploration has cleared significant forest areas and generated thousands of tons of waste; production has acidified and polluted waters, eroded lands, and polluted air; transportation has deforested more of the landscape; and refining has produced toxic sludge and additional water and air pollution.[27] Paradoxically, the transportability of oil compared with coal has launched the very technologies (pipelines, flow stations, chain saws, and so on) that have allowed oil to be tapped from deeper within the delta ecosystem, spreading the impacts more widely. In getting oil from the ground to tankers, over seven thousand oil spills, totaling between nine and thirteen million barrels, have seeped into the delta over the last fifty years, making the Exxon *Valdez* spill (0.27 million barrels) and the recent Gulf of Mexico disaster (nearly five million barrels) pale in comparison.[28] As we have learned from long-term studies of headline-grabbing spills, the impacts of oil persist for decades. For example, liver tests of ducks and sea otters continue to reveal hydrocarbons from the 1989 Exxon *Valdez* spill, oil from a 1979 spill in Mexico still contaminates the soil and mangrove forests there, and fiddler crabs in Massachusetts continue to suffer hydrocarbon-induced stupors from a spill in 1969.[29] With such a long-term outlook, it is believed that a financial valuation of impacts caused by oil activities in the Niger Delta would total tens of billions of dollars.[30]

There are additional, circuitous paths through which oil production has infiltrated the natural environment in the delta. As several of the artists in *Environment and Object · Recent African Art* make visible through works composed of plastic bottles and bags, petroleum-based products now also litter the landscape. Since organized waste management is nearly nonexistent in Nigeria, bags and bottles appear literally everywhere. Besides altering aesthetics, these materials pose significant danger to wildlife, clog pipes, and slowly decompose for hundreds of years, all the while releasing toxic fragments into the waters and soils of the region.[31]

The country's relationship to global climate change also acts somewhat as a double-edged sword and may prove the most damaging yet. While China and the United States are together responsible for over 40 percent of our global greenhouse gas (GHG) emissions, Nigeria holds a fairly high rank among the remaining countries (thirty-ninth in 2007).[32] The bulk of these emissions, however, results from gas flaring, deforestation, and other aspects of the delta's oil-export economy, and few Nigerians enjoy the luxuries that typically accompany significant GHG emissions. Yet the Niger Delta is predicted to experience a disproportionate cost of our global GHG emissions, with the most significant impact being a rise in sea level. Similar to the island nations that capture more media attention, the low-lying delta stands to suffer extensive erosion and inundation along more than 70 percent of the coastline, displacing 80 percent of the people and creating a severe impact on their major economic driver: oil.[33] It is a commonly held view that poverty begets environmental damage and environmental damage begets poverty, and, in the Niger Delta, this circular pattern will undoubtedly be challenging to break.

Oil production has also given rise to extreme occupational hazards in the delta, including countless deaths from explosions and fires and elevated levels of cancer and neurological disease,[34] but an outgrowth of the industry has also been violence. The environmental impacts of the expanding oil industry have largely destroyed rural livelihoods, and employment opportunities in the oil industry are few, even for young people who eke out a secondary and even a university education. Given these circumstances, youth have three options: migration, typically to the growing slums surrounding cities; remaining in the community and competing for very limited resources; or confronting the perceived enemy.[35] Relatively peaceful protests began in the late 1980s to draw attention to environmental degradation, indigenous rights, and the poor developmental state of the region, but the protests turned violent in the mid-1990s. More and more youth, provoked by disillusionment and frustration, sabotaged pipelines and oil installations, challenged community rule, and took hostages. A recent survey revealed that an astonishing 36.23 percent of the Niger Delta community members interviewed expressed a "willingness or propensity to take up arms against the state."[36]

The Niger Delta has truly become a zone of sacrifice economically, environmentally, and socially, and as one Niger Delta resident, Dorothy Ejuwa, gazes at the evening glare of a major oil refinery, she laments aloud, "For how long can we remain like this? That is our bitterness."[37] *Environment and Object · Recent African Art* rings with this warranted bitterness and makes clear the stranglehold of oil, but the exhibition also documents courage and empowerment. Many of the works express the inner strength of Nigeria, a commitment to turn waste once again into beauty, and a renewed call for questioning, justice, action, and change. In keeping with the export economy that has cost Nigeria so dearly, the exhibition engages the global community while asking us to explore and confront the ignorance that absolves us of collective responsibility.

1. Angus Maddison, *The World Economy: A Millennial Perspective* (Paris: OECD, 2001), 30.

2. J. R. McNeill, *Something New Under the Sun: An Environmental History of the Twentieth-Century World* (New York: Norton, 2000), 12.

3. Vaclav Smil, *Energy in World History* (Boulder, CO: Westview, 1994), 15.

4. McNeill, *Something New*, 15.

5. G. Tyler Miller, Jr., *Sustaining the Earth* (Pacific Grove, CA: Thomson, 2005), 202.

6. Country Watch, http://www.countrywatch.com (accessed August 30, 2010).

7. Michael Watts, "Sweet and Sour," in *Curse of the Black Gold: 50 Years of Oil in the Niger Delta*, ed. Michael Watts (Brooklyn: Powerhouse, 2008), 36.

8. McNeill, *Something New*, 304.

9. Watts, "Sweet and Sour," 43.

10. Central Intelligence Agency, *The World Factbook*, https://www.cia.gov/library/publications/the-world-factbook (accessed August 30, 2010).

11. United States Energy Information Administration, http://www.eia.doe.gov/ (accessed August 30, 2010).

12. Ibid.

13. National Intelligence Council, http://www.dni.gov/nic/NIC_home.html (accessed August 30, 2010).

14. Watts, "Sweet and Sour," 43.

15. United Nations Development Program, *Human Development Report 2009* (New York: Palgrave Macmillan, 2009).

16. Watts, "Sweet and Sour," 43.

17. Jeffrey D. Sachs and Andrew M. Warner, "Natural Resource Abundance and Economic Growth" (Cambridge, MA: Harvard Institute for International Development, 1997), 6–9.

18. Central Intelligence Agency, *The World Factbook*.

19. Sachs and Warner, "Natural Resource," 9.

20. Transparency International, http://transparency.org/ (accessed August 30, 2010).

21. Watts, "Sweet and Sour," 43.

22. Andre Kamdem Toham et al., "A Vision for Biodiversity Conservation in Central Africa" (Washington, DC: World Wildlife Fund, 2006), 6.

23. E. O. Wilson, *The Diversity of Life* (Cambridge, MA: Harvard University Press, 1992).

24. Paul Hawken, Amory Lovins, and L. Hunter Lovins, *Natural Capitalism* (New York: Little, Brown, 1999), 2–4.

25. Nenibarini Zabbey, "Impacts of Extractive Industries on the Biodiversity of the Niger Delta Region, Nigeria" (Nigeria: Center for Environment, Human Rights and Development).

26. McNeill, *Something New*, 304.

27. Federal Ministry of Environment, Lagos, et al., "Niger Delta Natural Resource Damage Assessment and Restoration Project: Phase 1—Scoping Report" (Washington, DC: World Wildlife Fund, 2006), 4–5.

28. "Gulf of Mexico Oil Spill," *New York Times*, http://topics.nytimes.com/top/reference/timestopics (accessed August 30, 2010).

29. Justin Gillis and Leslie Kaufman, "After Oil Spills, Hidden Damage Can Last for Years," *New York Times*, July 17, 2010.

30. "Niger Delta Natural Resource Damage Assessment and Restoration," 1.

31. John Roach, "Are Plastic Bags Sacking the Environment?" *National Geographic News*, September 2, 2003, http://news.nationalgeographic.com/news/ (accessed August 30, 2010).

32. United Nations, *Millennium Development Goals Indicators*, http://mdgs.un.org/unsd/ (accessed August 30, 2010).

33. Intergovernmental Panel on Climate Change, Fourth Assessment Report, http://www.ipcc.ch/; C. Okoko Ogba and B. Pius Utang, "Geospatial Evaluation of Niger Delta Coastal Susceptibility to Climate Change" (Sydney: FIG Congress, 2010, accessed August 30, 2010).

34. Santiago Borasin et al., *Oil: A Life Cycle Analysis of Its Health and Environmental Impacts* (Boston: Center for Health and the Global Environment).

35. Felix Tuodolo, "Generation," in Watts, *Curse of the Black Gold*, 114–115.

36. Aderoju Oyefusi, "Oil and the Probability of Rebel Participation Among Youths in the Niger Delta of Nigeria," *Journal of Peace Research* 45:4 (2008): 539–555.

37. Watts, "Sweet and Sour," 44.

Lisa Aronson and John S. Weber

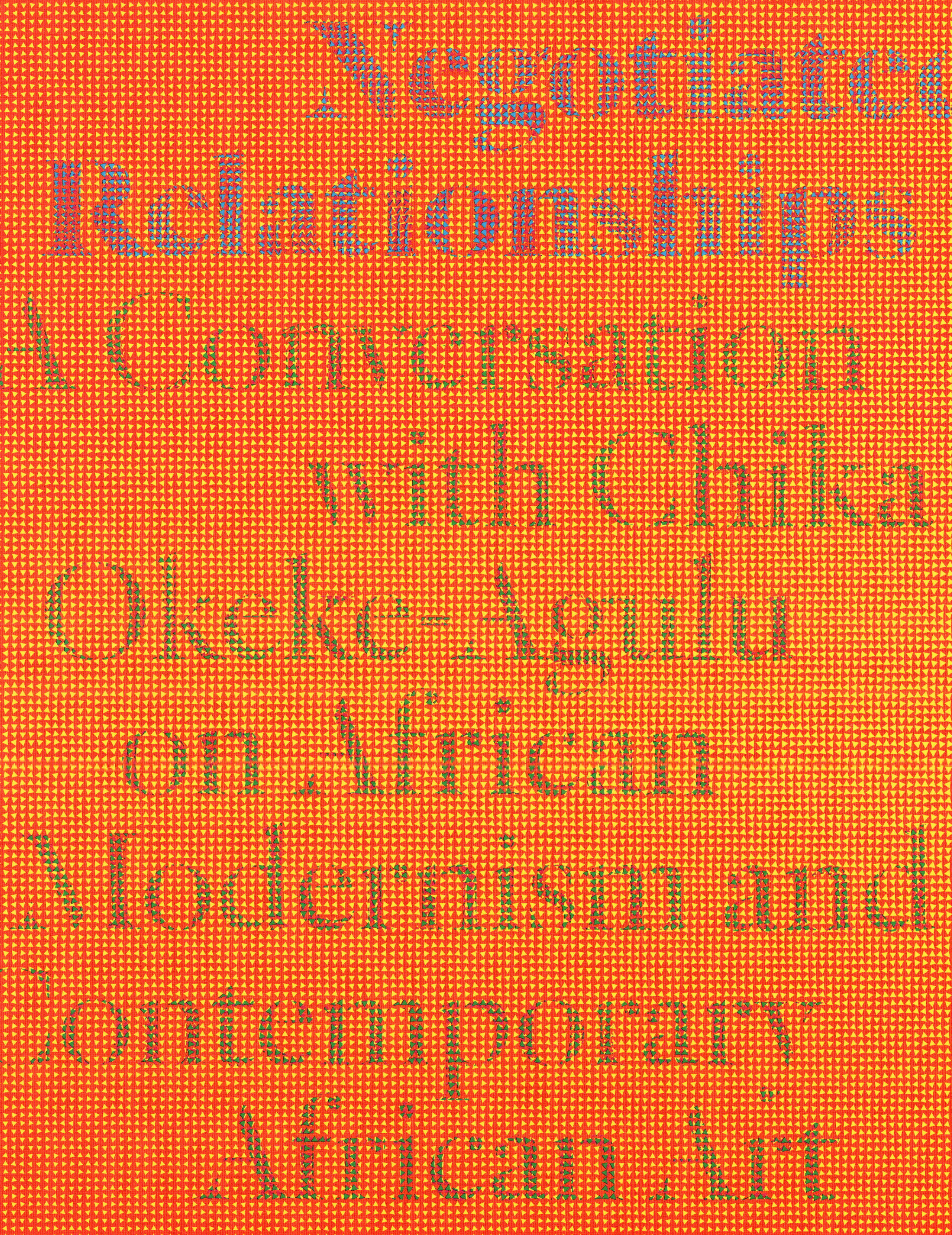

Negotiated Relationships

A Conversation with Chika Okeke-Agulu on African Modernism and Contemporary African Art

JOHN WEBER

frica is an immense continent with a dizzying range of people and cultures, a rich set of ancient histories, and many political and economic situations today. With this in mind, how do we begin to talk about contemporary African art? What, in your view, are its meaningful signposts, conditions, and characteristics? And is there a meaningful set of practices, assumptions, conditions, and shared heritages and situations that we can usefully refer to as contemporary African art, something, however varied, that is more than simply the sum output of Africa's many faces and places?

CHIKA OKEKE-AGULU

It's tricky talking about modern and contemporary African art, precisely because trying to demarcate at what point "the modern" and "the contemporary" begin is a hit-or-miss enterprise. We know that the sets of practices we describe as "modern" started at a particular point in time, the early colonial period, when artists came into contact with Western modes of artistic expression and representation. And then there are sets of practices and artists that came after that, leading up to the independence period. The question is at what point do we use the term *contemporary* to describe the work of African artists? This can be simply a semantic problem, but it can also be conceptual.

I would characterize "the modern" as sets of practices informed by the consciousness of decolonization—the kinds of artistic strategies and tactics that artists developed by the mid-twentieth century to deal with the consequences of independence and political sovereignty. What you found across the continent in the 1950s and the 1960s were artists grappling with what to do with their native and indigenous cultures, as modern artists. For me, this characterizes "the modern" because you see the development of avant-garde practices, artists trying to champion the new within their different countries. They were trying to demarcate boundaries with the past, while simultaneously negotiating the terms of their relationship with indigenous and Western traditions.

It's this dual track: a consciousness of the relevance of both the Western, which they learned in art schools or independently, and the indigenous, which they might have experienced or at least tried to reconnect to, that characterizes modern African art.

The "contemporary" implies not just an awareness or an attempt to negotiate one's relationship with tradition, but to go beyond that; it means to go beyond the national imagination; to connect with local but also global knowledge, information, and ideological networks. This is the argument that Okwui Enwezor and I make in our recent book, *Contemporary African Art Since 1980* (Damiani, 2009), although by the 1960s, such historians as Ulli Beier in his 1968 book *Contemporary Art in Africa* were already using the term *contemporary*.

Conceptually, what one refers to as contemporary African art indicates a clearly critical relationship with tradition, the nation, and the world. The way I think about it, contemporary African art can't simply be defined as, or indexed to, the fact that these artists were born in Africa or outside it; instead it's the relationship between this artist and these contexts—the nation, the continent, and one's place in the scheme of things globally—that is crucial. That's what defines my sense of contemporary African art. And this, for me, also reflects the incredible diversity of histories and experiences that constitute what we refer to as Africa today, in terms of the racial, demographic, and political diversities across the continent. So one has to then adopt a set of conceptual parameters to talk about what contemporary African art can mean.

J W

There was a huge expansion in contemporary art globally in the 1960s and 1970s, and then another big expansion that happened in the 1980s. And you're saying that for contemporary African art there's a fundamental conceptual shift that occurs, a shift in artists' relationship to tradition, and a more global context. What's the timeline?

C O - A

In the 1960s, there was a generation of artists—what I refer to as the independence generation of artists—that emerged from all across the continent. And this, of course, had a lot to do with the establishment of art schools in the 1950s and earlier. Along with their emergence was the establishment of critical forums like *Black Orpheus*, a journal of African and African American literature edited by Ulli Beier in Nigeria from 1957 through 1967, and the Mbari Artists and Writers Club that convened in Ibadan and other parts of Nigeria during the 1960s. The model was also adopted in East Africa in the Chemchemi Cultural Centre, the Paa-Ya-Paa Arts Centre, and so forth. So you had the emergence of critics—foreign and local critics—who were writing and creating a discourse around the work. And all this is responsible for the rise of this body of work in the 1960s, leading up to the *Contemporary African Art* exhibition at London's Camden Arts Centre in 1968–69, which remains the largest exhibition of modern and contemporary African art ever in terms of the number and diversity of artists and in terms of the volume of work exhibited. That Camden Arts Centre exhibition was the summation of the modern period.

What happened in the 1970s was a moment of questioning. And you see that emphatically in Dakar, with the rise of the Laboratoire Agit-Art to challenge the Senghorian model of modern art. You also see it in East Africa, where people challenged the Margaret Trowell pedagogy, and at the University of Nigeria at Nsukka, where a new school emerged. So in many parts of the continent there was this moment of reengagement with what it meant to be contemporary artists working in Africa.

But this reengagement also was checked by overall disillusionment across the continent, because in fact the kinds of critical networks established in the 1960s pulled back in the 1970s, with the collapse of the Mbari-Chemchemi networks in Nigeria and in East Africa. The Paa-Ya-Paa was still running, but the range of connections possible during the independence period of the 1960s all but ended. *Black Orpheus* ceased publishing in 1968 (it was revived later, but as a shadow of its former self); *Transition* magazine went into crisis in East Africa and reemerged in the United States.

And so the decade of the 1970s was a moment of pause, as Okwui Enwezor and I argue in our book. This is not to say that artists stopped making work, but when you look at the collective, continent-wide discussions in the 1960s about what it meant to be artists—and writers were doing the same thing—you had very little of that anywhere on the continent during the 1970s.

By the 1980s, new groups were appearing, whether in the Sudan, with the Crystalists; or in Dakar, where the Laboratoire Agit-Art became much more robust by the early 1980s; or in Nigeria where the Aka Circle of Artists became the most important artists' group during the 1980s and 1990s; or the Vohou-Vohou group in Sierra Leone; or the Sisi-Kwa-Sisi movement in Kenya. We can track the rise of the careers of many African artists to these and similar groups talking among themselves and trying to reconnect not just with other groups within the country but also outside of their countries. The London-based Triangle Arts Trust, whose workshops were established in many African countries, played a pivotal role in fostering these new networks. In fact, the grounds for the current "successes" in the careers of African artists were already being established in the 1980s, even if their work only began to attract global attention in the 1990s with the big exhibitions and biennials.

Of course, these timelines are arguable in terms of the development of what we are looking at today, but this timeline is one that I want to stick with.

Would you say that the 1980s was a more globalized moment than what happened in the 1960s, the artists of the independence movement?

I'm not quite sure if the independence moment was less global. What I'm more certain about is the reception of the work of artists in the 1960s, and that of the artists in the 1980s and especially the 1990s. It's not as if the earlier generations of artists weren't showing overseas. Christian Lattier, Ibrahim El Salahi, Skunder Boghossian, Uche Okeke, Jimo Akolo, Malangatana Ngwenya, Twins Seven-Seven, and many others were showing in Europe and the United States. But if you look at the venues of their shows, you see that commercial gallery or art museum exhibitions were rare for these artists; it was normal in the United States to present their work in libraries, community centers, and historically black art spaces, which indicates the ideological context of the work's reception. I doubt that Ben Enwonwu, arguably the most internationally famous modern African artist ever, had a show in any major commercial gallery in the United States, and that says a lot in terms of the reception of modern African art in the West. In Europe, things were a bit different, as many of the artists showed in commercial galleries, but their works remained marginal or were even ignored in discussions about contemporary art there.

The other thing is that the Harmon Foundation, which was initially charged with the promotion of African American artists, was responsible for the exhibitions of most of these African artists in the United States. Although the Harmon played a tremendous role in the careers of black artists, its collaboration with African artists meant their work could not escape being mediated through the discriminatory infrastructures of American racial politics. In Europe, African artists' work was—at least initially— presented as evidence of colonialism's vaunted civilizational mandate: exhibitions of African artists were taken as proof that the project of colonization was yielding fruit, because their works showed signs of modernity, but with an asterisk.

This is different from what happened after the 1989 *Magiciens de la Terre* at the Pompidou Center in Paris, Grace Stanislaus's *Contemporary African Artists: Changing Traditions* at the 1990 Venice Biennial, and Susan Vogel's *Africa Explores* show at New York's Center for African Art in 1991. Beyond that, you had the *Seven Stories About Modern Art in Africa*, which I co-organized at the Whitechapel Art Gallery in 1995, and Yukiya Kawaguchi's *An Inside Story: African Art of Our Time* at the Setagaya Art Museum

in Tokyo, also in 1995. These shows took place for the most part in important modern
and contemporary art museums, and the venues of exhibitions say a lot about the
reception of the art and the artists. All of this is to say that artists were going out of Africa
in the 1960s and before, but their work was received in a particularly marginalized
cultural context and shown in spaces quite different from those enjoyed by the artists of
the 1990s and after.

J W

**I'm curious about travel, residence, and the back-and-forth movement of the liberation
generation on the one hand and the post-1980s generations on the other.**

C O - A

The question about what happens to artists in the 1990s and after, in terms of the
colonial affiliations of the previous generation, is a very interesting one. It gets to the
heart of the difference between the post-independence modern period and the
1980s and beyond.

In the 1960s—or even before the 1960s, during the colonial period—artists mostly went
to the metropolitan cities of the colonizing countries to train or, in the case of South
Africa, to escape racist and oppressive sociopolitical conditions. After the 1960s, you
had a lot more art schools in Africa, reducing the need to train overseas. But also the
landscape of contemporary art changed. Paris and London—especially London—ceased
to be as attractive, with New York, Amsterdam, and Berlin getting into the mix.
Paris's allure never quite diminished, and this no doubt reflects the fact that France still
maintains strong political and economic influence over its former colonies to this day.
So Francophone African artists still go to Paris, whereas the center of Anglophone
African art and scholarship has no doubt become the United States. Once the formerly
inflexible lines of colonial allegiance dissipated, artists became a lot more aware
of more diversified spaces and locations of international contemporary art and took
advantage of that diversity.

Having said that, in terms of important exhibitions of contemporary African art
outside the continent, Europe is still the place, at least from my own experience,
while most non-exhibition-related scholarship comes out of the United States. If you
look at the major and canonical exhibitions, only Susan Vogel's was produced in
the United States. *The Short Century: Independence and Liberation Movements in Africa,
1945–1994* (2001) originated in Germany; *Seven Stories* (1995) came from London;
while *An Inside Story* (1995) and *Africa Remix* (2004) were from Japan and Germany,

respectively. So, very clearly, the Europeans still invest a lot more in producing important exhibitions of contemporary African artists, even though the strongest markets and intellectual networks are in the United States.

That brings us to the exhibition you recently curated in Berlin at the Nationalgalerie. How did you select the artists, and what does it mean to have such a big project devoted to contemporary African artists in the capital city of a major European nation?

The project, *Who Knows Tomorrow*, was initiated in 2007 by the now former president of Germany, Horst Köhler. And there is a lot to be said about the fact that the president of a European nation actively supported the production of a major exhibition of African artists' work. The initial expectation was that we—myself and Britta Schmitz—organize the usual group exhibition of African artists to accompany a conference that was part of the president's Africa Forum initiative on African economic problems. We decided not to do a group exhibition and were supported by Udo Kittelmann, who joined the curatorial team once he became the director of the Nationalgalerie.

We wanted to step out a bit from the typical large-but-important group exhibition format and invite just a few artists to occupy historic art museums in Berlin, not only the capital but a city going through tremendous transformation itself. We wanted conceptually profound, formally ambitious, spectacular work, because we wished to make a statement about contemporary African art and about colonial history and the postcolonial condition. As much as one would like to believe that, yes, contemporary African art has arrived, many remain skeptical; so this provided a good opportunity for an emphatic statement about the achievement of contemporary artists from the African continent.

The five artists we selected— El Anatsui, Zarina Bhimji, Antonio Ole, Yinka Shonibare, and Pascale Marthine Tayou—have a remarkable record of engaging with the link between colonization and the postcolonial condition of Africa and Europe. And, of course, Berlin was the venue for the Berlin-Congo Conference (1884–85), where Africa was divided up, under the auspices of German chancellor Otto von Bismarck, the political ancestor of President Köhler who initiated our project. And so we wanted to connect the practice of these artists to this political and historical context. Of course, we could have included several other artists, but again, that would then have led to this football-team idea of representation; so we stayed with just five artists who presented their individual installations in four museums of the Nationalgalerie.

You said that by bringing these artists together you were making an emphatic statement. Could you pursue that further?

It is to announce again that contemporary African artists are making some of today's most important art. And that's why I noted that it had to be spectacular—not for its own sake, but to show that these artists are dealing with very complex formal questions in their work, and even more complex and profound conceptual issues.

Do you see yourself as an ambassador for African art?

I don't see myself as an ambassador. I see myself as an intellectual activist. I'm more comfortable thinking of myself in those terms. Through our work as historians, critics, and curators, my generation wanted to confront directly the realization that African artists were doing important work but weren't given a chance to be seen and heard. And it has been rewarding when you look at what has happened to the field of contemporary African art and see how far it has come within a relatively short period of time.

In view of that, let's look at the question of the inherited tendency of African art historians and curators to approach African art anthropologically and approach it through the lens of traditional African art. There's definitely a tension there between earlier generations of scholars and curators, and historians or critics with a contemporary focus.

Quite a bit of the earlier scholarship tended to look beyond the work of art and to talk about the artists and the "African" cultures that supposedly produced them. There was a greater urge to link these artists to some indigenous African arts and cultures in order to prove that African cultures are alive. So there was a premium on the work of artists who are purportedly advancing the art of their ancestors, artists in whose blood an inexorable ancient African river flowed. The model of this scholarship was colonial cultural anthropology, which often ended with a paragraph in which the author briefly explained how the studied culture has changed in the face of modernization. But in most of these analyses you didn't see a deep engagement with the work of art, with its formal and conceptual conditions. The work tended to be taken as evidence of a social dynamic, rather than as the product or process of an intellectual and creative imagination. The emergence of modern and contemporary African art history, I believe, rested and still depends on putting the artwork at the center of a critical discourse rather than using it as an excuse to analyze changes in modern African cultures and societies.

Having said that, I want to track back to relate this with the work that artists did in the 1960s in a concerted effort to mine their indigenous traditions as resources for their art. This connects to what I was describing earlier as an important characteristic of African modernism—the artists' negotiation with what they considered their indigenous traditions. But they negotiated as part of a critical poetics and ideological rhetoric, not as a way to *continue* or update their ancestral traditions, an intention the scholarship in effect imposed on the art. So there was a disconnect, a misalignment, between artists who were imagining a critical relationship with African traditions, and anthropological scholarship keen to demonstrate how artists were continuing, for example, their Yoruba or Igbo or Zulu arts. You still see this in some recent scholarship, most blatantly in the ways that some critics and museums are struggling to make El Anatsui's work intelligible by inscribing it solely within the traditions of West African indigenous arts. Of course, anthropology ought to continue doing what it does best, but such examinations cannot be the grounds of a sustainable and robust contemporary African art history or criticism.

How about the relationship to tradition and history in terms of where artists live today? One of the things we have grappled with in this show is how you define an "African artist" at all, given how mobile this generation of artists has become.

One of the more interesting recent issues for contemporary African art is this question of whether artists live inside or outside of the continent. It's almost as if we're back to the question of authenticity, a big issue in traditional African art—you know, "was this object used for actual rituals or not?"

The question of authenticity has taken a different guise now: if you are an African artist living and working inside the continent, you're supposedly more authentic than one who lives elsewhere. But just as with the earlier authenticity issue, this one is quite reductive and actually unhelpful for how we think and talk about the work of artists from the African continent. The lives of people on the continent today— what they know, what they see, what they hear—are influenced, as Arjun Appadurai once argued, by external and internal forces, movements, new technologies and information networks. They look to the United States, to Europe, to Japan, to Saudi Arabia, to Dubai and to other African countries. They look at the past with an eye to the present and future in all their confounding complexities. So you are talking, actually, about a global condition experienced from particular fixed or shifting locales.

The other thing to consider is what happened to many African countries after the imposition of the International Monetary Fund's Structural Adjustment Program in the early 1980s. You have to appreciate the resulting devastation of the continent's economies and its educational and political institutions, and the ensuing massive brain drain, to understand the choices contemporary artists make about where they want to live and work. In the same way that many of their peers in the sciences, social sciences, and humanities went overseas when it became impossible to do meaningful work at home, many artists relocated to keep abreast of the work of their peers and to gain access to the networks of art—galleries, museums, journals, and so forth.

I sometimes think of these artists and scholars as migrant workers. The migrant worker goes elsewhere to work in order to support the family back home, and he often goes back and forth between the land of birth and the place of work. Think of an artist like Barthélémy Toguo, who has just built the Bandjoun Station, his art center, in Cameroon. He lives in Europe, but he couldn't possibly have done what he has done in his home village today if not for his work's European successes. And his work continues to deal with sociopolitical issues affecting the continent in much the same way that it did before he moved to Europe, although naturally his public is much wider now that he has access to galleries, museums, and fairs all over the world. Many artists like him live in the West—think of Kendell Geers, Georges Adeagbo, Pascale Marthine Tayou— but have strong connections with people in their native countries, while their work derives from and speaks to their experience of home and the world. On the other hand, the people they left behind—extended families, friends, and others—depend financially on them. So the inordinate emphasis on the African artist's place of residence, whether she lives inside or outside the continent, speaks to a misunderstanding about why people (including artists) decide where to live and work, whether by choice or compulsion.

Their practice also reflects, in true fashion, what we mean by globalization and the global condition, which can make where you live important only to the extent to which it affects the life and career opportunities available to you. What's more important to me as a scholar is how any art or artist responds to ideas and discourses crucial to our understanding of contemporary art in its immediate and extended horizons. I am less interested in where artists pay their taxes or which passports they carry, because it may have little or no bearing on the work artists do or how they imagine themselves as human beings.

The show we're working on concerns the environment, broadly defined. Could you talk about how you approach the question of the environment as it affects contemporary art and African artists working today?

The whole environmental movement today, I would argue, has not become important in the work of African artists. And I think this also has to do with the question of environmental awareness. It's still nascent across the continent, except in the more drastically degraded environments, such as the Niger Delta, which you now see in the work of photographers like George Osodi and the Depth of Field group in Lagos. The incredible degradation of the environment, because it's apparent everywhere, has elicited these artists' critical response. But for the most part, environmentalism has not become prevalent in the work of contemporary African artists.

Today you do still see landscape paintings—say in the work of the Yaba School in Lagos—where it's more about celebrating the environment, from the northern Nigerian landscape to the chaotic and colorful Lagos metropolis. This approach is quite different from the work of artists taking a more critical position on the natural and social environment. Someone like Bright Eke, for instance, started making sculptures and installations with plastic bottles and water sachets that people discard in the streets, and inevitably makes us conscious of urban environmental pollution, even as he transforms the wastes into dramatic, even sensuous installations.

There's also another aspect to this, evident in the work of the Vohou-Vohou group in the Ivory Coast in the 1980s and the early 1990s, but also in that of the Ona artists at Obafemi Awolowo University in Ife, Nigeria. Both groups focused on using colored soils, organic substances, and locally available materials. They rejected standard art materials—marble, oil, acrylics—and looked to their own environment for media and techniques of art making. These artists began to wonder, "Why spend money on important materials when they could do without them?" This came, of course, at a point when—in the wake of the 1980s Structural Adjustment Programs—economic conditions in Africa had soured and whatever institutional support there was for art had more or less disappeared. Although they were responding to socioeconomic realities, their reimagined work became in itself an ideological enterprise, that is to say, the reclamation of the local, the rejection of the foreign, and a critique of the effects of neocolonialism. In any case, the work these artists did conveys the nature of the environment and the kinds of resources available in that environment.

You spoke of the oil burning in the Niger Delta. Can you contextualize that?

People in the United States were seized in 2010 by the fever and horror of the Deepwater Horizon oil spill in the Gulf, and you couldn't do anything without hearing news or seeing the video of the gushing oil. The entire system scrambled to fix the spill. But this has been going on for decades in the Niger Delta. There, oil leaks, spills, and gas flaring are a way of life for the local people. What has happened to the Niger Delta is so outrageously terrible that it's gaining the attention of contemporary artists and photographers working in Nigeria—George Osodi, for instance, but also Uche James Iroha, and his Lagos-based Depth of Field colleagues. One can think of their work as environmentalist in the sense that it calls attention to this terrible thing being done to the environment there.

As you see in the photographs of George Osodi, the Niger Delta looks like a lunar landscape. The vegetation in many parts is almost completely decimated. These people live on fishing but now can't fish anymore in the lifeless, polluted creeks and waterways. The situation there is much worse because there is active collusion between the Nigerian government and international oil companies and foreign governments to ignore or play down the extent of the devastation of the Niger Delta. Think of the judicial murder of Ken Saro-Wiwa, the writer and leader of the Movement for the Survival of Ogoni People who was executed in 1995. He didn't deserve to die, but the oil companies and the General Abacha dictatorship thought sacrificing him would stop the rising voices in the Niger Delta. Fortunately, his death didn't completely silence those critical voices. Moreover, groups like the Environmental Rights Action in Nigeria led by Nnimmo Bassey have been working incredibly hard to bring what is going on in the Niger Delta to the attention of the world.

Of all the artists that we've selected, Jerry Buhari seems to embrace the landscape image most traditionally. I'm wondering where he got that from and why you think he continues to pursue it.

One important development in contemporary Nigerian art that concerns the environment revolves around the work of two artists associated with the Ahmadu Bello University in Zaria: Gani Odutokun (1946–1995), and Jerry Buhari, who is in your show. Zaria has had a long-standing tradition of landscape painting. In the late 1950s and 1960s, most of the Zaria students came from southern Nigeria; as such, the semi-arid Zaria environment was a novelty to them and captivated their artistic imagination. In the paintings, you see these Zaria landscapes with traditional Hausa architecture and people wearing flowing gowns, the *baban riga*.

Gani Odutokun and Jerry Buhari brought to bear on this Zaria tradition an ethical approach to the landscape. Buhari's liquidized oil paintings from the early 1990s, for instance, addressed environmental degradation. At the time, importation of oil paints was virtually banned, since the dictatorship considered them a nonessential commodity, and artists grappled with ways of finding cheap, locally made art materials, which had also become quite expensive. So when Jerry Buhari decided to pour oil paint on the canvas, basically letting it waste away, he was doing something besides painting. This was his way of calling attention to the wasting of valuable resources in Nigeria—oil and natural gas, for instance, not to mention the vast human resources. On the face of it, this very formal enterprise, reminiscent of the Abstract Expressionist gestures of someone like the American Morris Louis, was in fact a clearly articulated argument for an environmental ethics.

J W

You have said that contemporary African artists and art share none of the anxieties of "the modern" that characterized the liberation generation of artists. What has changed?

If you imagine the burden that these artists had in the years immediately after independence—defining and leading the newly independent nation to greater heights in order to show the colonialists "Yes, we can do it"—then it is easier to understand the anxiety I am talking about. In what direction do we lead the nation? In what language shall we write? Do we write in our native languages because that's what most of our people read? Or do we write in the colonial languages, which are only accessible to the literate few, though we are trying to align ourselves with the masses of our people? Artists were working under the tremendous stress of representing and emblematizing everything progressive about the new nation. But also, as I stated earlier, they confronted the question of the place of traditional and ancestral arts in their own work.

What you find in the work of later generations is a decision, conscious or otherwise, not to be saddled with these questions. And I think that is not unrelated to what happened to the nations themselves. In Nigeria, the civil war broke out in 1967, following two military coups and mass killing of the Igbo living in other parts of the country in 1966. As a result, most of the Igbo artists and writers previously living and working in Lagos alongside their other Nigerian colleagues were forced to pull back, and, in fighting for the short-lived Republic of Biafra, developed a much more critical attitude to the idea of Nigeria, even after the war. You see similar disenchantment with the postcolonial national project in other countries, though not always induced by civil war. In Senegal, the Laboratoire Agit-Art developed around the critique of the Senghorian negritude ideology. In the Congo (formerly Zaire), Mobutu Sese-Seko outlined the idea of *authenticité* as the basis of his own nationalist project and required all citizens to change their French and Belgian names to their more authentic ancestral Congolese names. Yet you have artists like Tshibumba Kanda-Matulu, and Chéri Samba who, through their so-called "popular art," challenged the government-imposed national reorientation and thus echoed popular disenchantment with what obviously were excesses of the postcolonial state.

You see similar kinds of critical positions in the work of artists from various parts of the continent. It is as if the nation were no longer an entity to be defended and chaperoned as in the 1960s but an idea to be critically refined, its assumptions and modalities subject to debate and critique. Invariably the idea that they were representing their nations became largely irrelevant. From the 1980s onward, artists became less concerned with anchoring their identity in the postcolonial nation. Instead, they think for the most part in terms of their own practices as individuals who live and work in a particular locale, yet are part of a larger global community.

PRINCETON UNIVERSITY

JUNE 30, 2010

John S. Weber

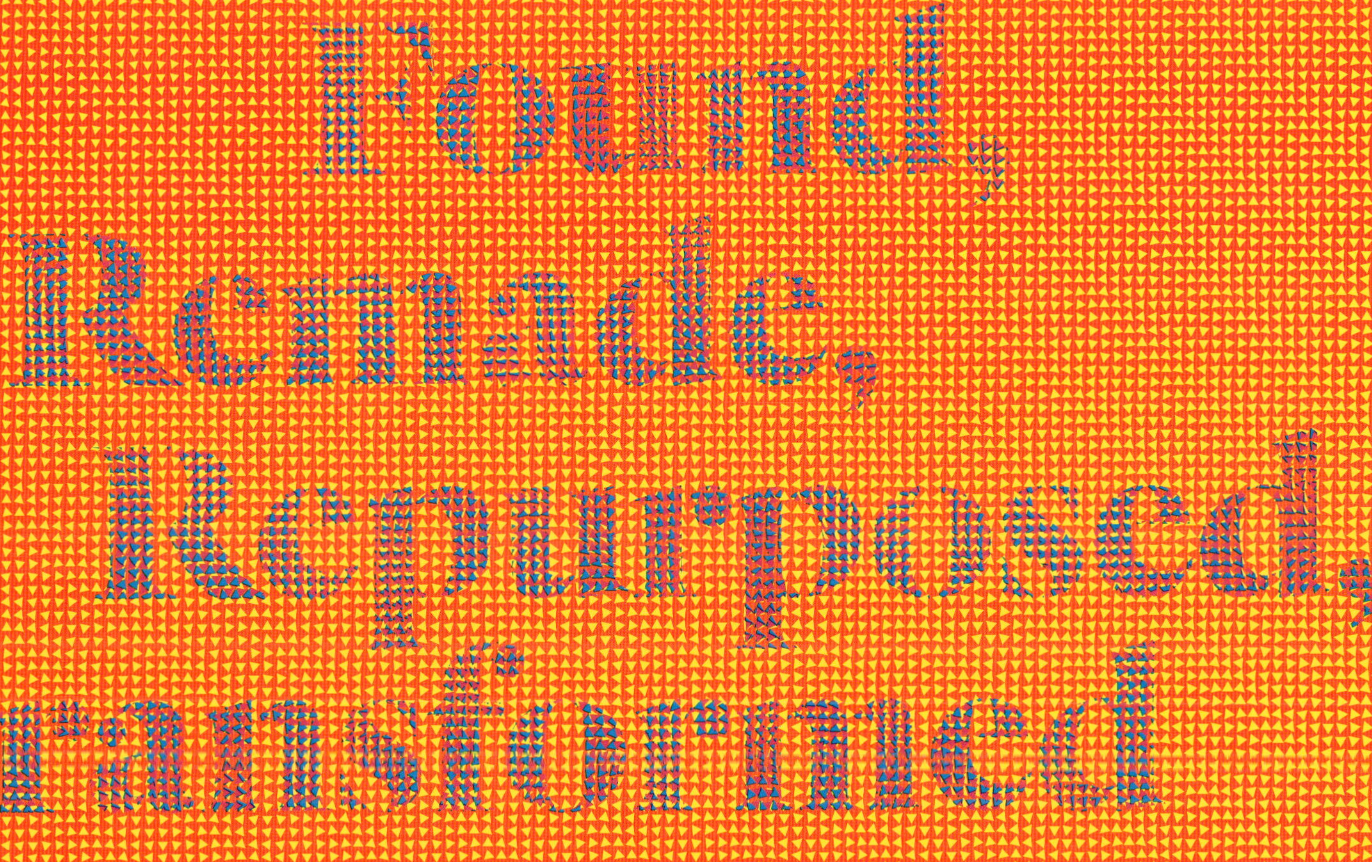

the Ghana-born, Nigeria-based artist **El Anatsui** has often spoken eloquently about the central role of found objects in his creative process. Over the course of his forty-year career, the practice of making art from things that "the environment has thrown up," to use his terminology, has increasingly shaped the meaning, form, and reception of his internationally celebrated body of work. Indeed, his influence as a teacher and artist shows clearly and undeniably in *Environment and Object · Recent African Art*. Yet the practice of claiming, repurposing, and renegotiating the meaning of found objects is common to many contemporary artists from Africa today, just as it has long been a staple of modern and contemporary art globally. Found objects bring their histories and contexts with them into works of art, and in *Environment and Object* artists take up found materials to signal a range of ecological, political, social, and historical concerns. In their hands, mundane objects also achieve unexpected and breathtaking beauty, and this, too, is on full view in the exhibition.

El Anatsui's spectacular works made from used and flattened bottle caps and metal-foil bottle tops exploded onto the international art scene in the first decade of the twenty-first century. His impact reached a crescendo in 2007 at the Venice Biennale with a cluster of astoundingly beautiful pieces anchoring the corner of the Arsenale, while another magisterial work draped across the façade of the Palazzo Fortuny. Amid the historical layering and the at times splendid decrepitude of Venice, El Anatsui offered a vision of luxe and excess fashioned from scraps of garbage and insignificant refuse. His work stopped exhibition visitors in their tracks, their mouths open, heads shaking, amazed that something so glorious could be fashioned from objects so trivial and debased. Since then, new exhibitions and catalogues have testified to the impact of El Anatsui's work, digging into his long history and its many distinct periods and bodies of work.

It is a measure of El Anatsui's originality that there remains even now no accepted term to describe the artist's signature works. The phrases "metal wall sculptures," "wall hangings," "sculptural wall hangings," "metal canvas," and simply "sculptures" all appear in the catalogue of his recent retrospective at the Museum for African Art in New York. Likewise, the possible relationships of these works to the *kente* cloth of his country and region of origin have been both celebrated and disputed. The question of El Anatsui's relationship to African textiles reflects long-standing clashes in the field of African art history, between anthropological approaches seeking to establish connections between traditional arts and modern or contemporary artists and those of contemporary art historians claiming that essential distances and departures separate past and present African artists. Arguably, the capacity of El Anatsui's work to excite such controversy, attracting and sustaining many, often competing discourses, represents the most convincing testimony to its art-historical significance and ultimate durability.

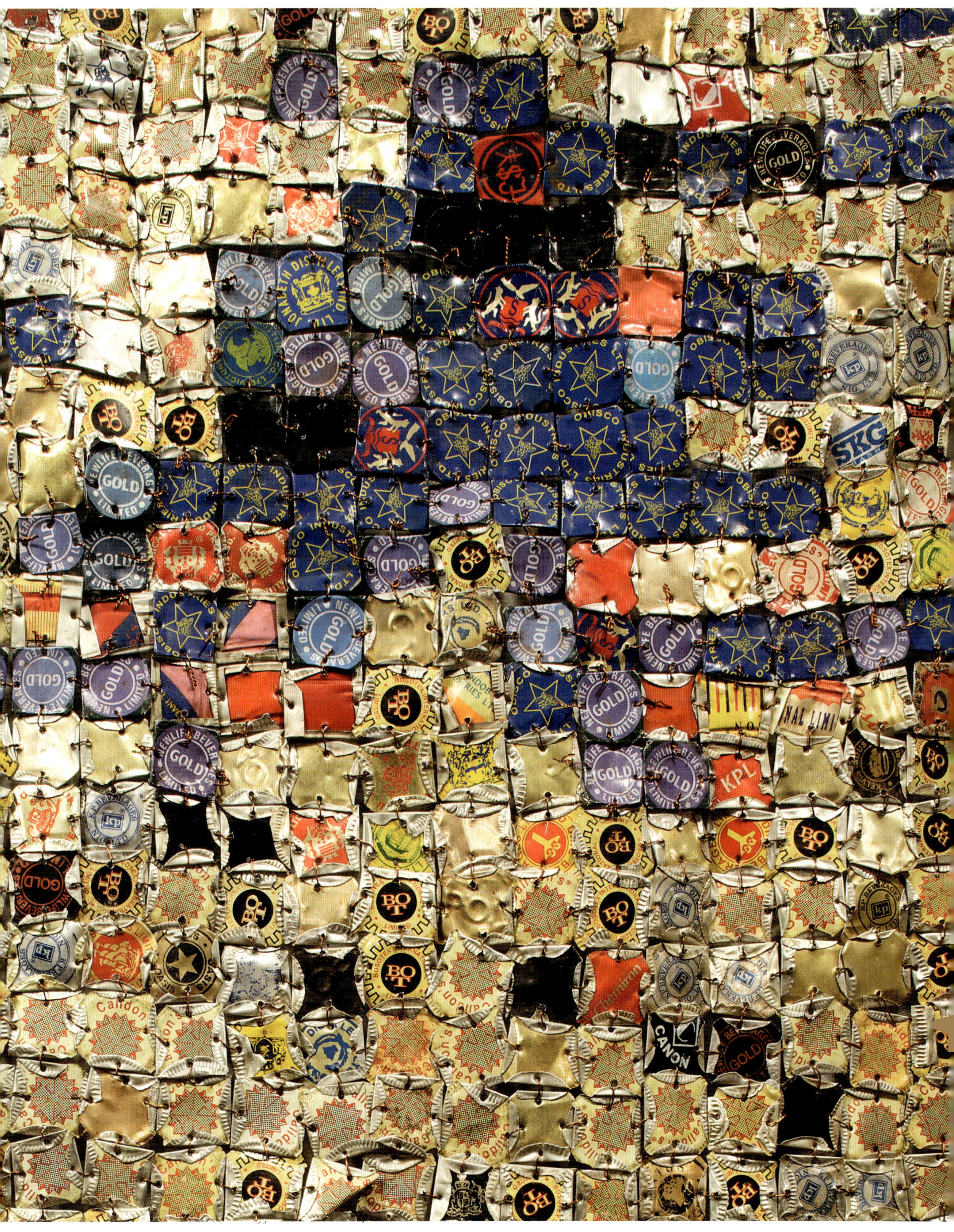

EL ANATSUI
Some Still Come Back, 2005
Aluminum and copper wire
92½ × 119 × 6½ inches

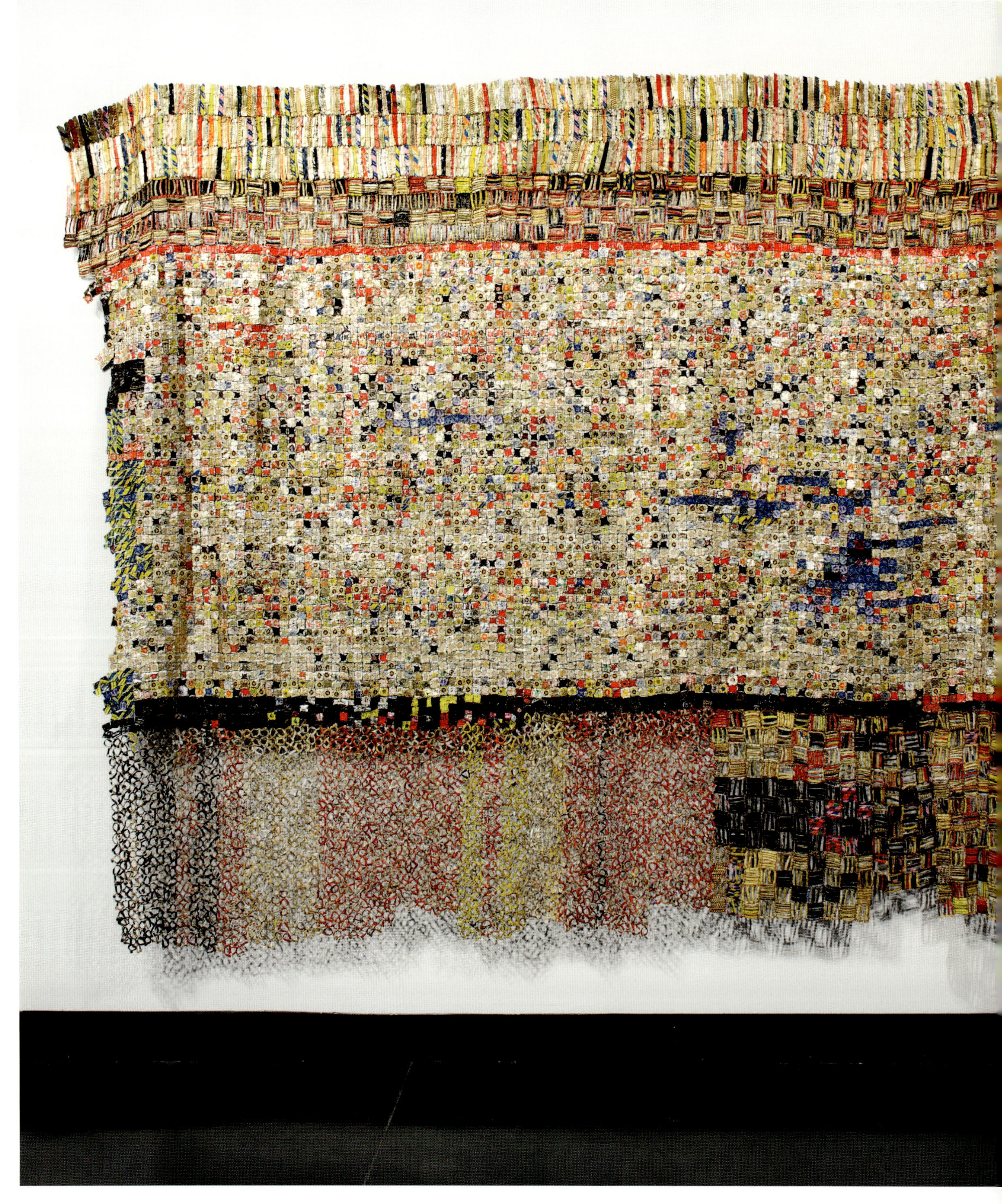

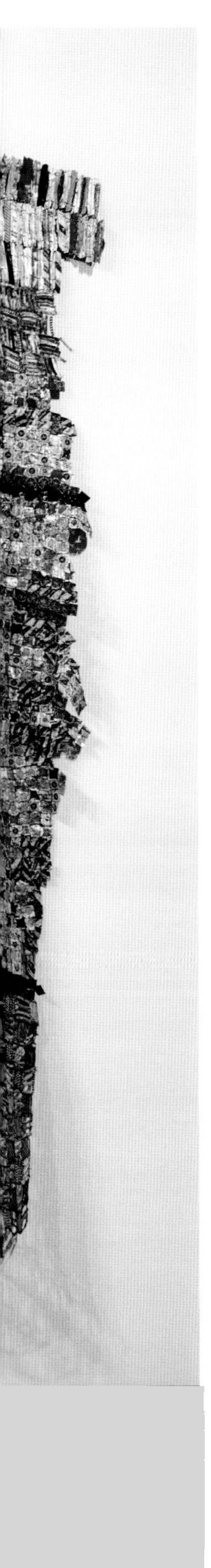

Speaking of his original impetus to create wall works from metal scraps, the artist has said:

Several things went through my mind when I found the bag of bottle tops in the bush. I thought of the objects as links between my continent, Africa, and the rest of Europe. Objects such as these were introduced to Africa by Europeans when they came as traders. Alcohol was one of the commodities brought with them to exchange for goods in Africa. Eventually alcohol became one of the items used in the transatlantic slave trade. They made rum in the West Indies, took it to Liverpool, and then it made its way back to Africa. I thought that the bottle caps had a strong reference to the history of Africa.[1]

At other times, he has noted that he prefers to work with "something that has its own story or its own history,"[2] and "things which have been used before, things which link people together. If you touch something, you leave a charge, and anybody who touches it is connected with you."[3]

In these and other comments, El Anatsui demonstrates a historically layered and critical relationship to his materials. Readily acknowledging that his own biography has equipped him well to appreciate the relationships of his celebrated work to Ewe *kente* cloth, Anatsui nevertheless speaks with greater frequency and engagement about the industrial nature of his materials and their relationship to larger issues of history and sociopolitical environments. In one interview, he even likens the process of making his sculptural hanging pieces to working with digital media, since he builds each work up one metal "pixel" at a time into segments that he then combines to make a whole piece. "There are a lot of decisions to make—it's like working with digital media—you have a section which you don't find OK, you can easily lasso it and pick from another place to replace it."[4]

Visually, Anatsui's metal hangings run the gamut from the huge, virtually architectural interventions of Venice and his 2010 Berlin exhibition to much smaller, domestically scaled pieces. Some offer stylishly patterned designs featuring large areas of one color shot through with broad stripes, while other works, such as *Some Still Come Back* (2005) from *Environment and Object*, offer a grittier, less immediately ingratiating mix of surface texture and overall pattern. Some works sprawl or cascade onto the floor, while others hover slightly above it. The artist disavows a strictly prescribed approach to the presentation of his pieces and characteristically asks collectors and curators to determine for themselves precisely how to hang his work. Regardless of how they are installed and named, El Anatsui's recent metal-sculpture wall hangings rank among the most acclaimed and admired examples of art being produced today.

Two other artists featured in *Environment and Object*, **Bright Ugochukwu Eke** and **Nnenna Okore**, demonstrate a commitment to the use and deployment of found materials that bears witness to their years of study with El Anatsui and attests to the productive shadow he casts. Yet both artists also stake out new terrain in choosing and using their materials.

Bright Eke created *Ripples and Storm 1* and *2* on site at the Tang Museum in only three-and-a-half weeks, working with recycled bottles, fallen trees salvaged from the Skidmore College Northwoods, and a small crew of students and professional installers. Together, *Ripples and Storm 1* and *2* embody the fluid and overlapping boundaries between notions of "environment" and "object" that motivated the exhibition. Eke's installation functions equally as an ecological critique of issues relating to water use, plastics, and waste management and also as an energizing exemplar of art created from found objects. As such, *Ripples and Storm* proved the perfect visual introduction and conceptual primer for *Environment and Object*.

Despite a previous site visit, Eke began the process of creating a new, site-specific installation at the Tang without a precise design or plan beyond the loose notion of fashioning the image of a "storm" that would hang in the museum's atrium and Illumination Gallery. His working process involved sorting the bottles according to color, piercing them in a uniform manner, then stringing them together with wire to form a pliable set of large prefabricated components that he could manipulate to compose the final work. Initially thinking that he would intermix the wooden "coins" cut from the tree trunks and singed by a blowtorch, Eke ultimately chose to use them in a pendant work installed alongside the bottle piece.

Eke speaks of water as his medium, but he treats it only as a conceptual and thematic starting point for architec-tural-scale installations rich in visual drama and broad metaphoric resonance. In earlier pieces, such as *Acid Rain* (2005), Eke took up explicitly ecological themes, and re-cent works, such as *Ripples and Storm*, continue to reflect

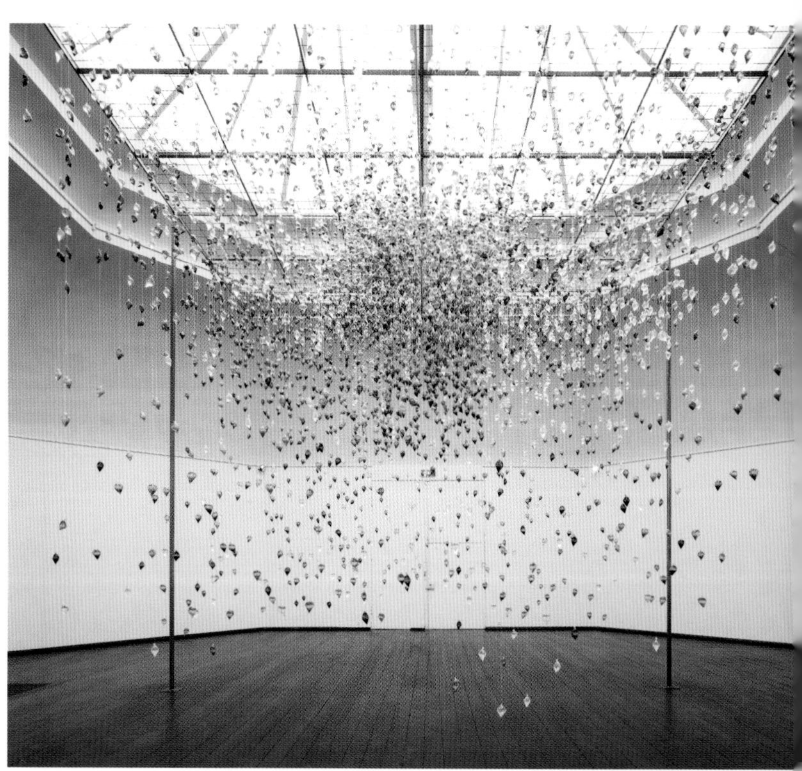

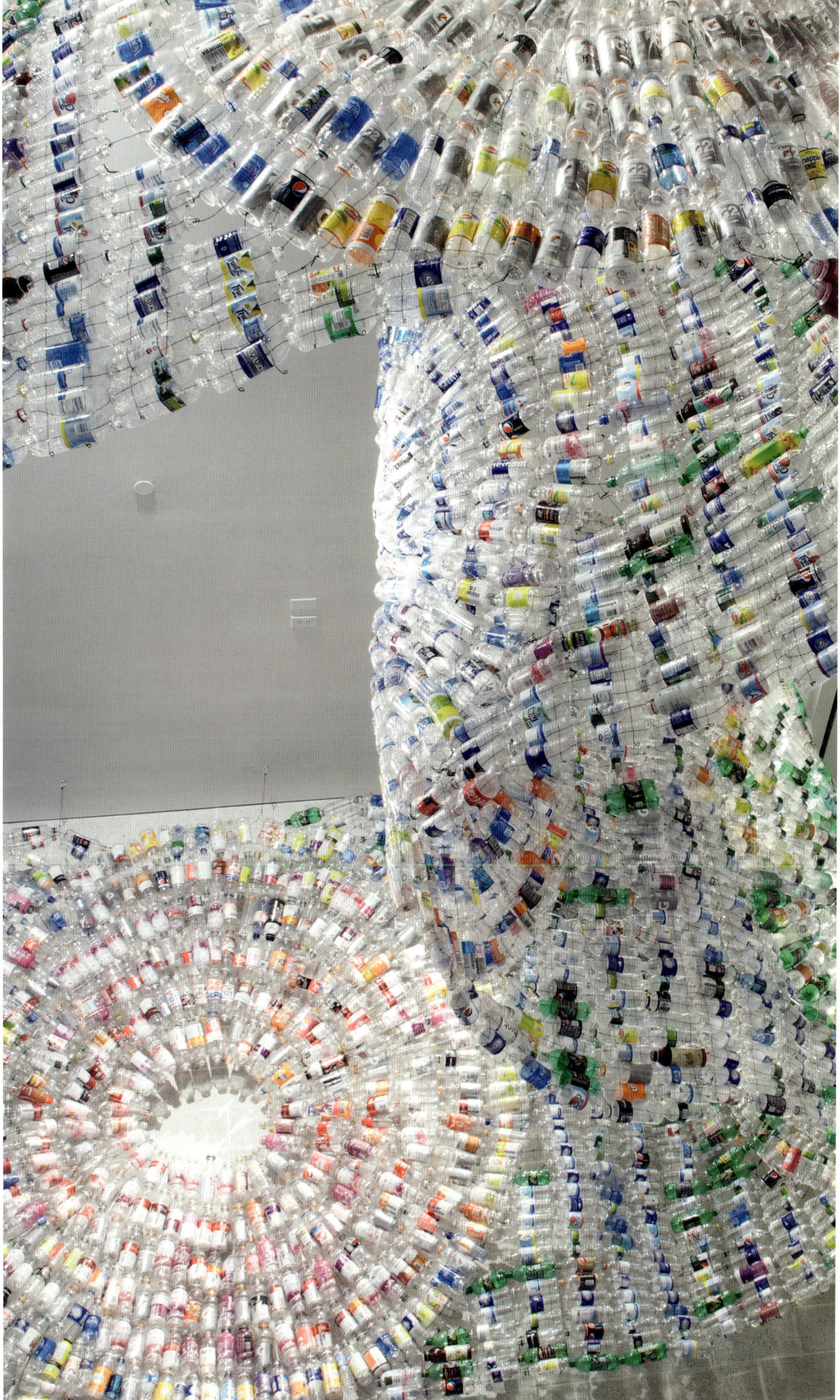

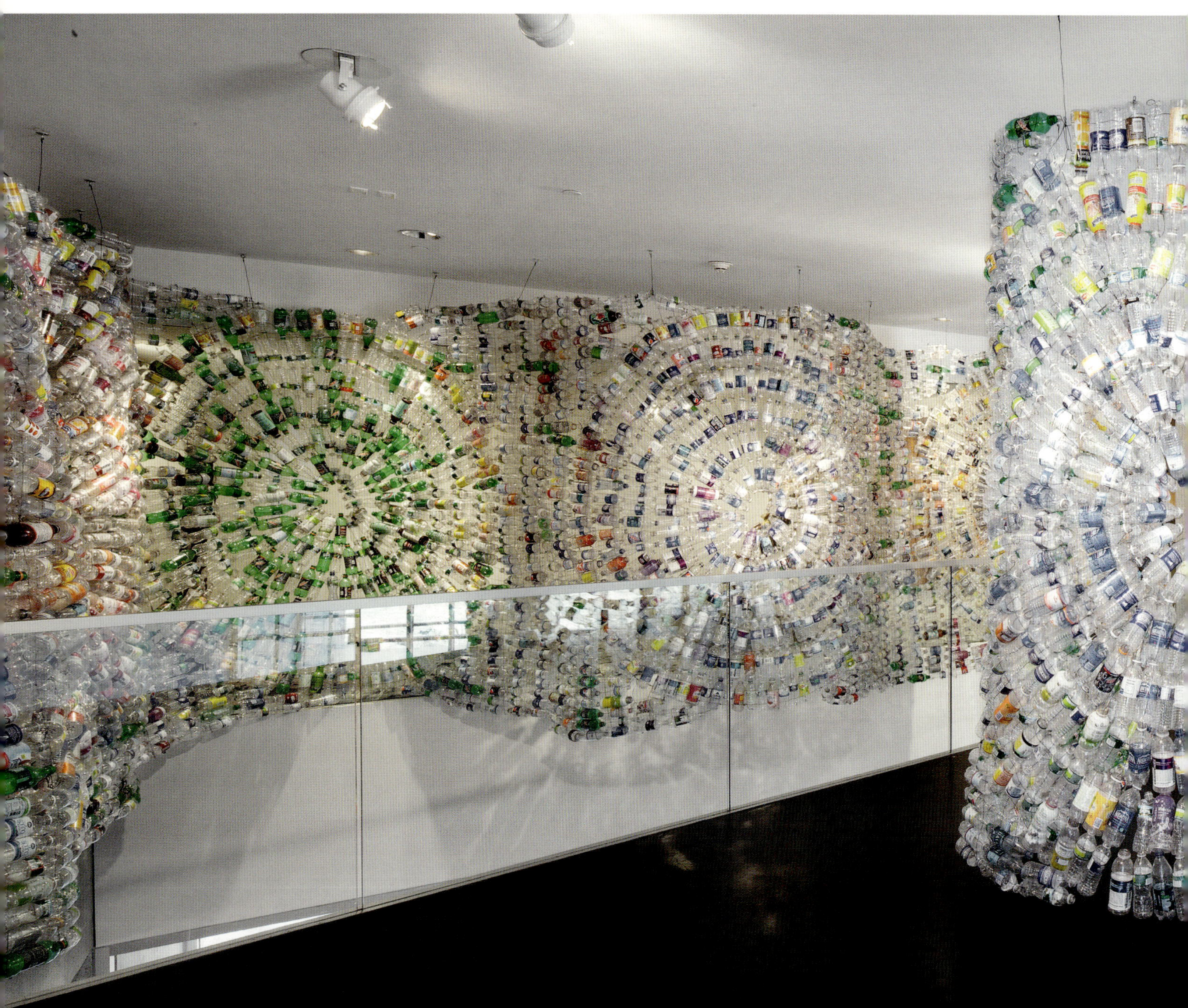

BRIGHT UGOCHUKWU EKE
Ripples and Storm 1
and **2**, 2011
Recycled bottles, wire,
salvaged wood, and
hardware
Installation dimensions
variable

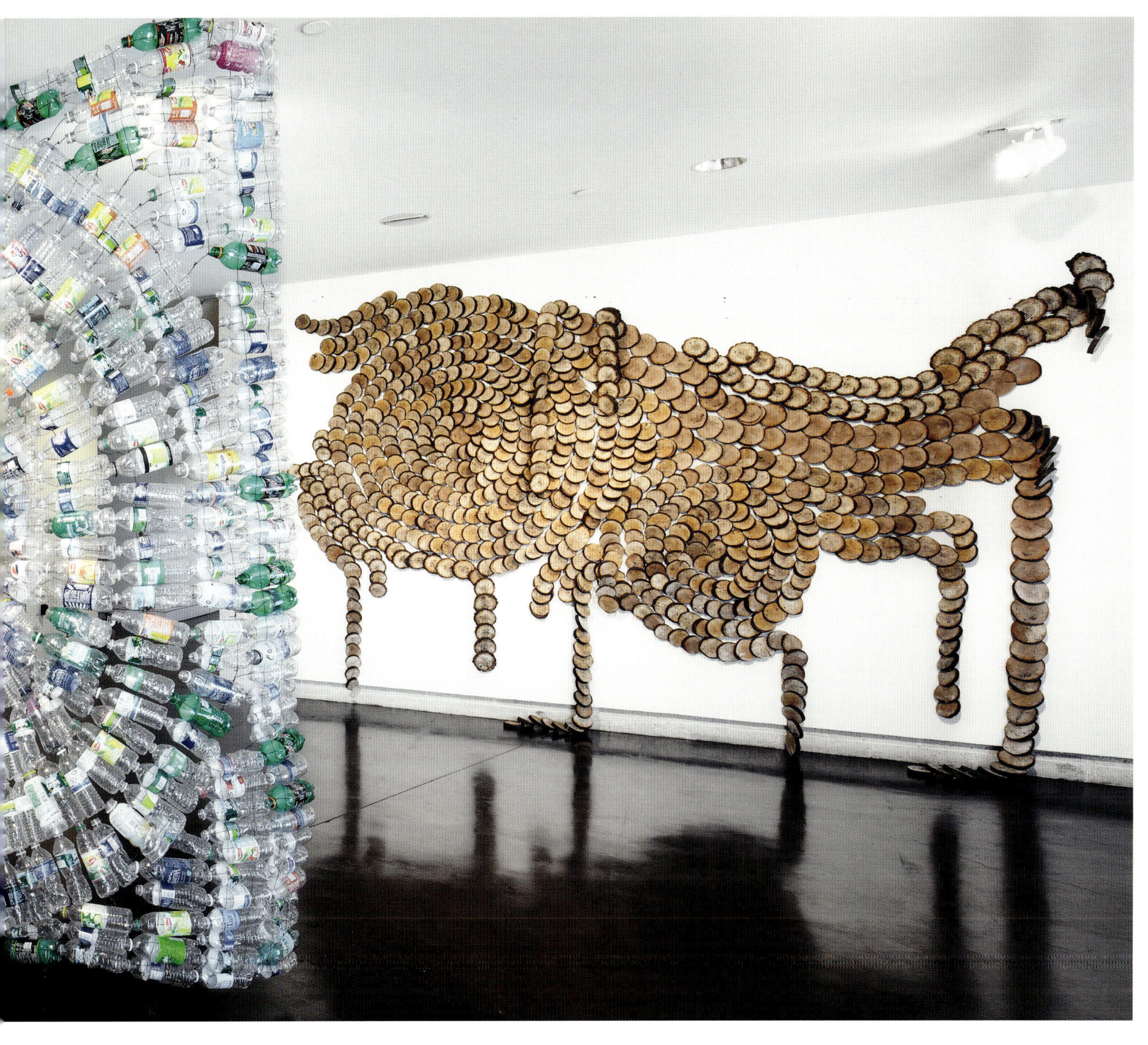

John S. Weber

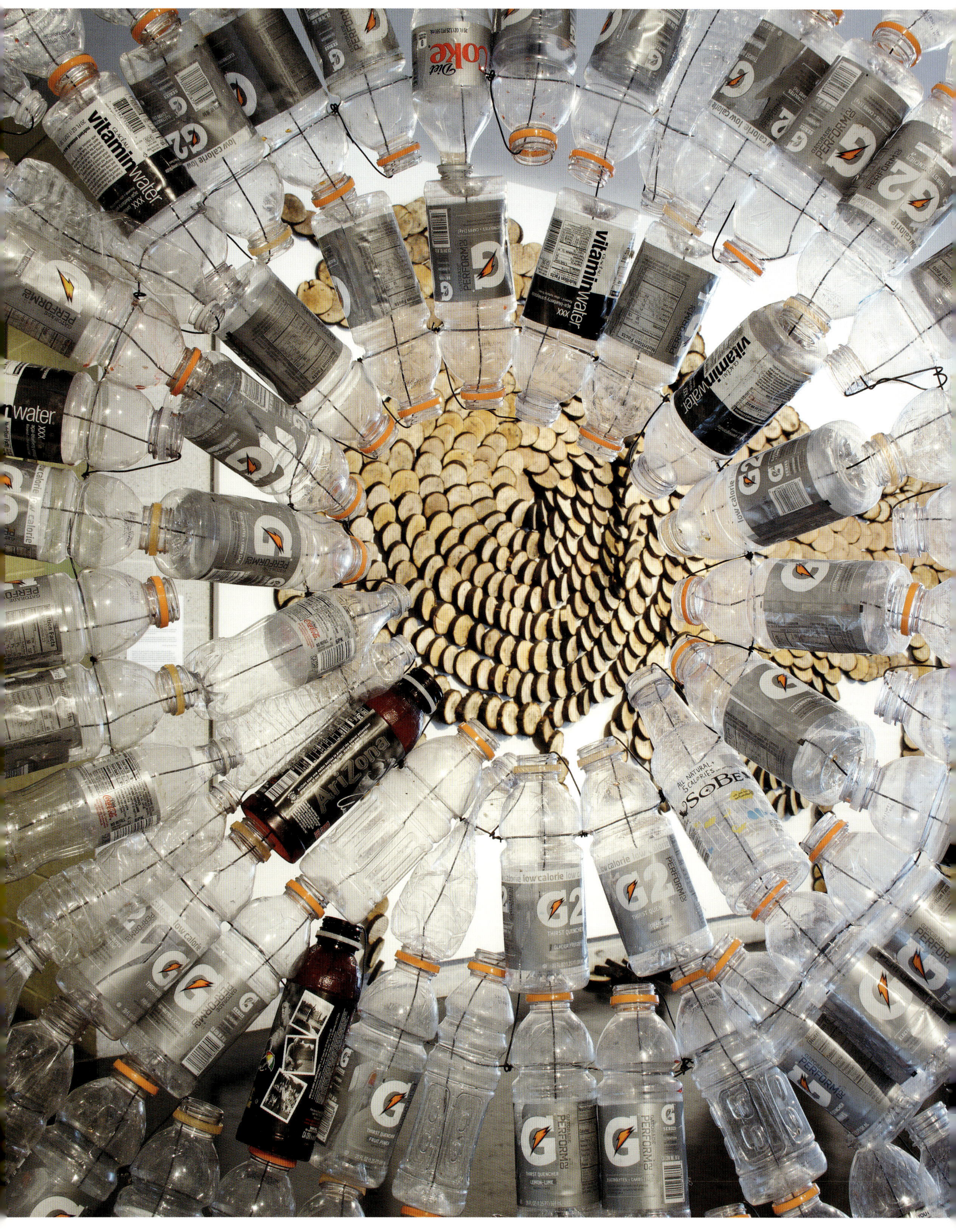

his environmental awareness. Viewers savoring the formally and experientially resplendent, sinuous walls of *Ripples and Storm* will be hard pressed to forget that they are seeing thousands of discarded plastic bottles destined for the landfill or, at best, a recycling bin. The work contrasts beauty with trash and overproduction even more explicitly and shockingly than El Anatsui's work, even as both artists' capacity to transform ugliness into visual pleasure remains formidable.

Wary of being cast as simply an ecological protester, Eke hastens to identify himself as an artist interested in environmental issues and not as an "environmental artist" or activist. Eke also notes the importance of transparency in his bottle installations, which create walls and spaces that allow viewers to see one another yet still keep them apart. His pieces need to be experienced bodily as well as visually, and his wooden wall piece, *Ripples and Storm 2*, rewards viewers' movements by offering a warm but shifting mix of charcoal and wood tones. The wooden wall piece also provides a stark contrast with the overwhelmingly "artificial" feel of Eke's bottle installation, complicating notions of the relative merits and aesthetic appeal of natural and organic materials. Yet for the artist, both pieces reflect the reality of our dependence on water as the essential source of all life and the basis of the ecosphere itself.

Nnenna Okore's sculpture and installation works have used wood, clay, wire, textiles, and both found and handmade paper, and she has also worked in video and photography. *Mbembe* (2009), her piece in *Environment and Object*, employs one of the most ubiquitous and unassuming materials imaginable: used plastic bags. Tying predominantly white bags in a loose web, she has fashioned an oversized net that hangs from the wall and drapes onto the gallery floor.

The pale color and soft texture of the bags give *Mbembe* a strangely fluffy, almost snowy aspect, warmed by the regular appearance of red Target logos and a few yellow and black "smiley" faces printed on a number of the bags. The overall effect is at once magical and mundane. The eye savors the subtle play of light across *Mbembe*'s lacey, ropey topography, while the mind mulls the idea of a work of fine art constructed solely out of this throwaway material of the cheapest order. Okore conjures up an atmosphere of

    John S. Weber

NNENNA OKORE
Mbembe, 2009
Plastic bags
Dimensions variable,
approx. 137 × 153 × 110
inches

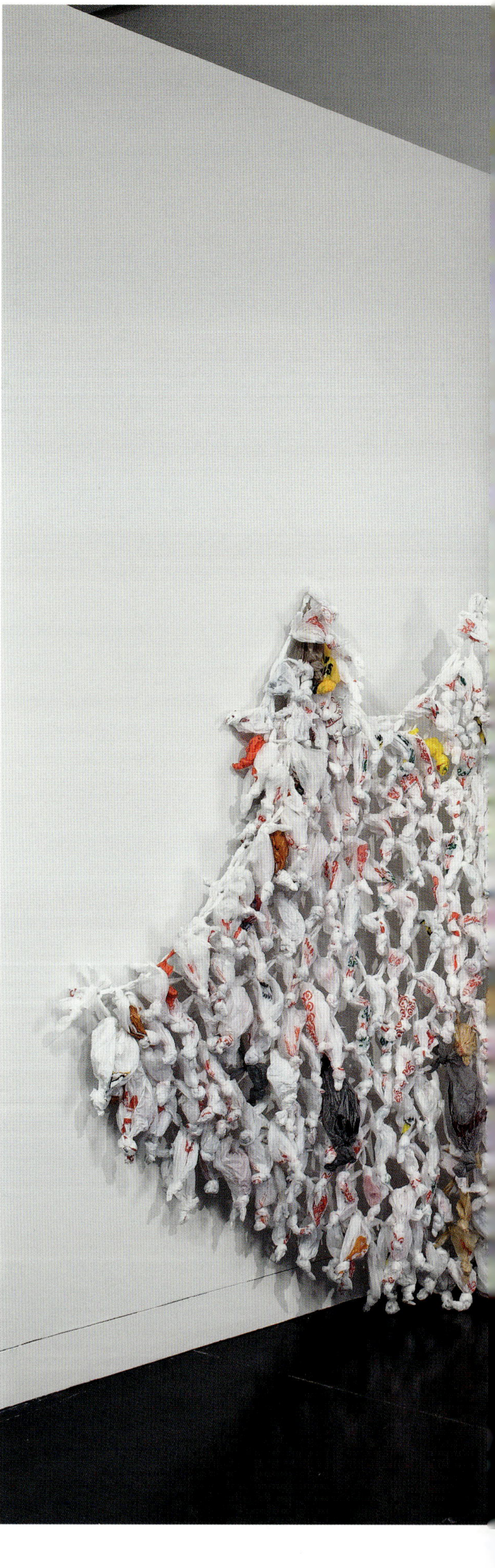

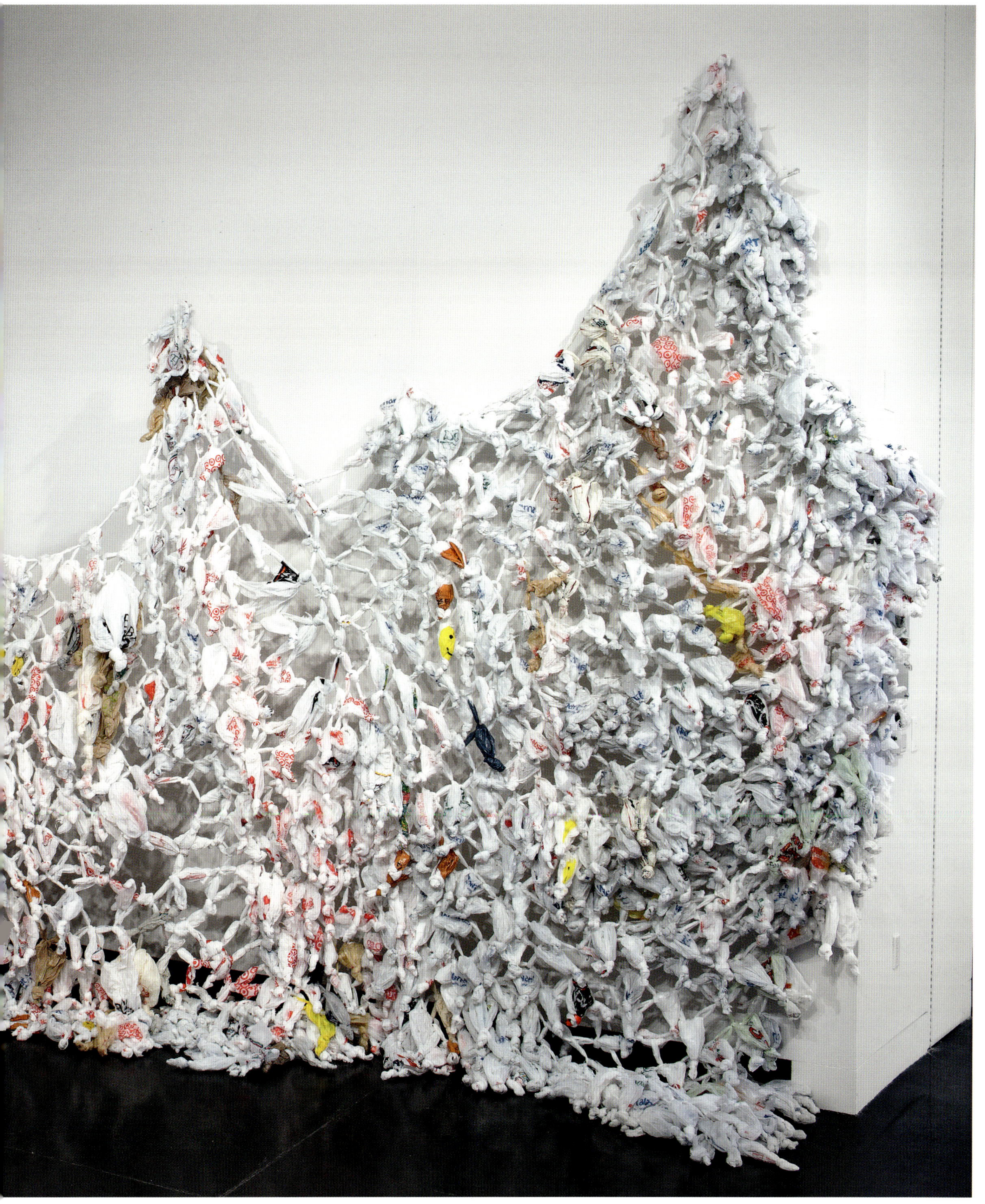

     John S. Weber

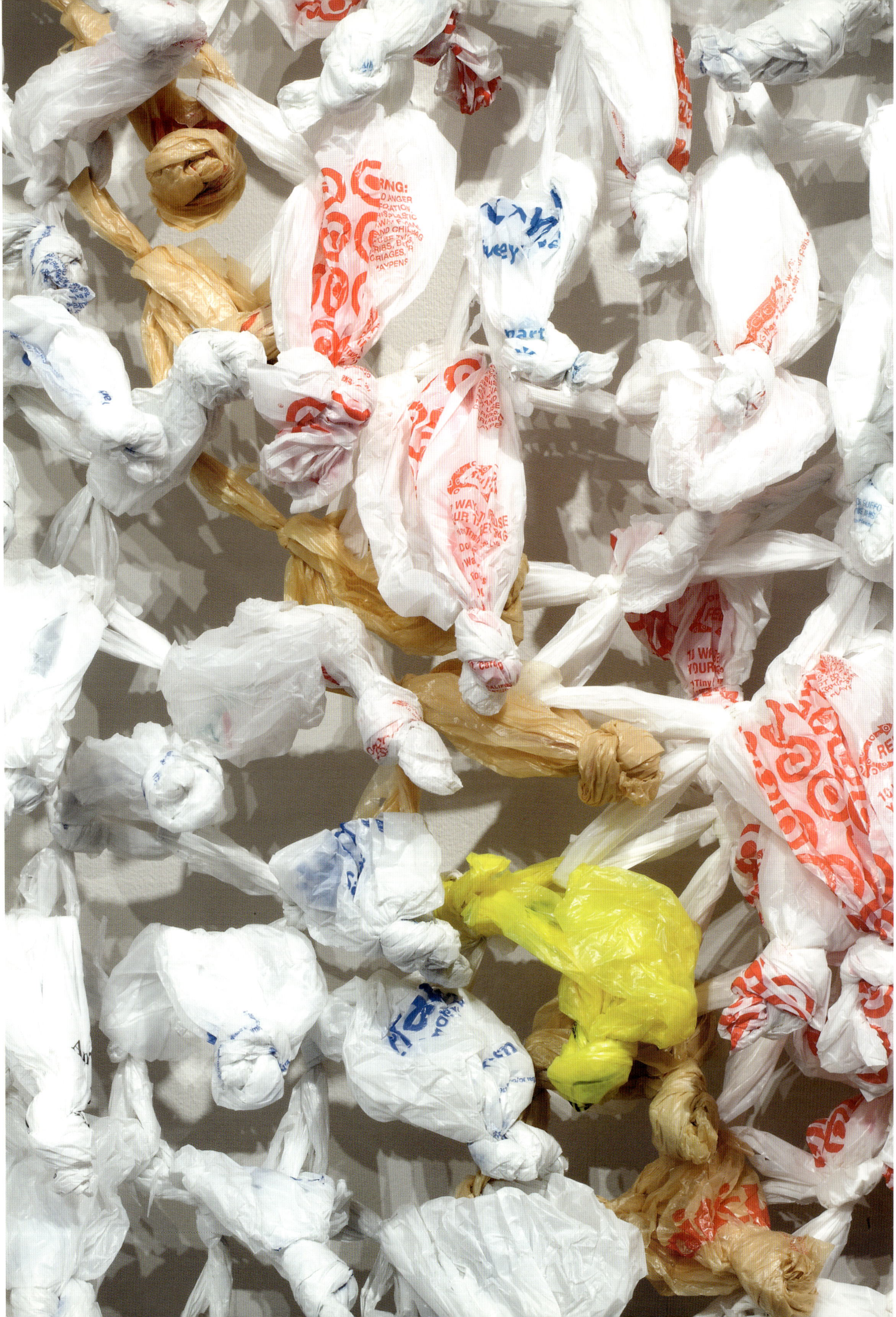

plush excess, yet her installation ultimately poses a sharp question about the economic and ethical values of a consumer society that can blithely fill those bags and then throw them all away. The title, *Mbembe*, means "shadow" in Igbo, Okore's first language. According to the artist, it suggests the way that discarded plastic bags have become an unintentional shadow cast in the wake of industrialized cultures. Mostly non-biodegradable and unrecycled, they follow us everywhere, signaling the things we have purchased, the places we've been, and our relentless pace of consumption.

Reflecting Okore's experience of the American economic environment in which she has lived and worked for a number of years, *Mbembe* also echoes the practice of reusing everything possible, which she absorbed during her African childhood. She has written of her fascination with the ingenuity villagers demonstrate in creating new value out of discarded objects, precisely the methodology on view in *Mbembe*.

Plastic bags would seem an unusually poor base material for fine art, yet the unexpected alchemy they undergo in *Mbembe* lies at the heart of Okore's work, as well as that of Eke and El Anatsui. In each case, the process transcends "recycling" and becomes transformation, a point El Anatsui has made. To his mind, recycling involves returning something again to its original function, that is, remaking a substance to perform again as it originally had. Paper remains paper, glass remains glass, plastic remains plastic. Such works as *Some Still Come Back*, *Ripples and Storm*, and *Mbembe* gather, organize, and deploy a multitude of individually insignificant, industrially manufactured products, fundamentally transmuting them into something unforeseen and unintended by their original producers.

Transformation also plays a role in the work of **Romuald Hazoumé** of Benin, but here it operates in a distinctly historical register. While many contemporary African artists tend to reject the impulses of art historians and anthropologists to read their work against a backdrop of traditional African arts, Hazoumé deliberately recalls traditional Africa in his series of updated masks made from used jerry cans and other modern objects. His work often possesses a humorous edge, but it is a sharp edge that cuts politically against any tendency to see his masks as exercises in mere historical irony or visual cleverness. This incisiveness becomes strikingly clear in such pieces as *Tchin-Tchin, BP!* (2010), a sculpture fashioned from dark, worn, and abraded jerry cans stapled together in a skinny column that flares at the top and bottom, roughly mimicking the shape of a champagne glass. As with his masks, Hazoumé positions and cuts the plastic cans so that their handles and short spouts unmistakably evoke a set of haunted faces, seemingly singing or chanting. In a statement for *The Global Africa Project* exhibition at

     John S. Weber

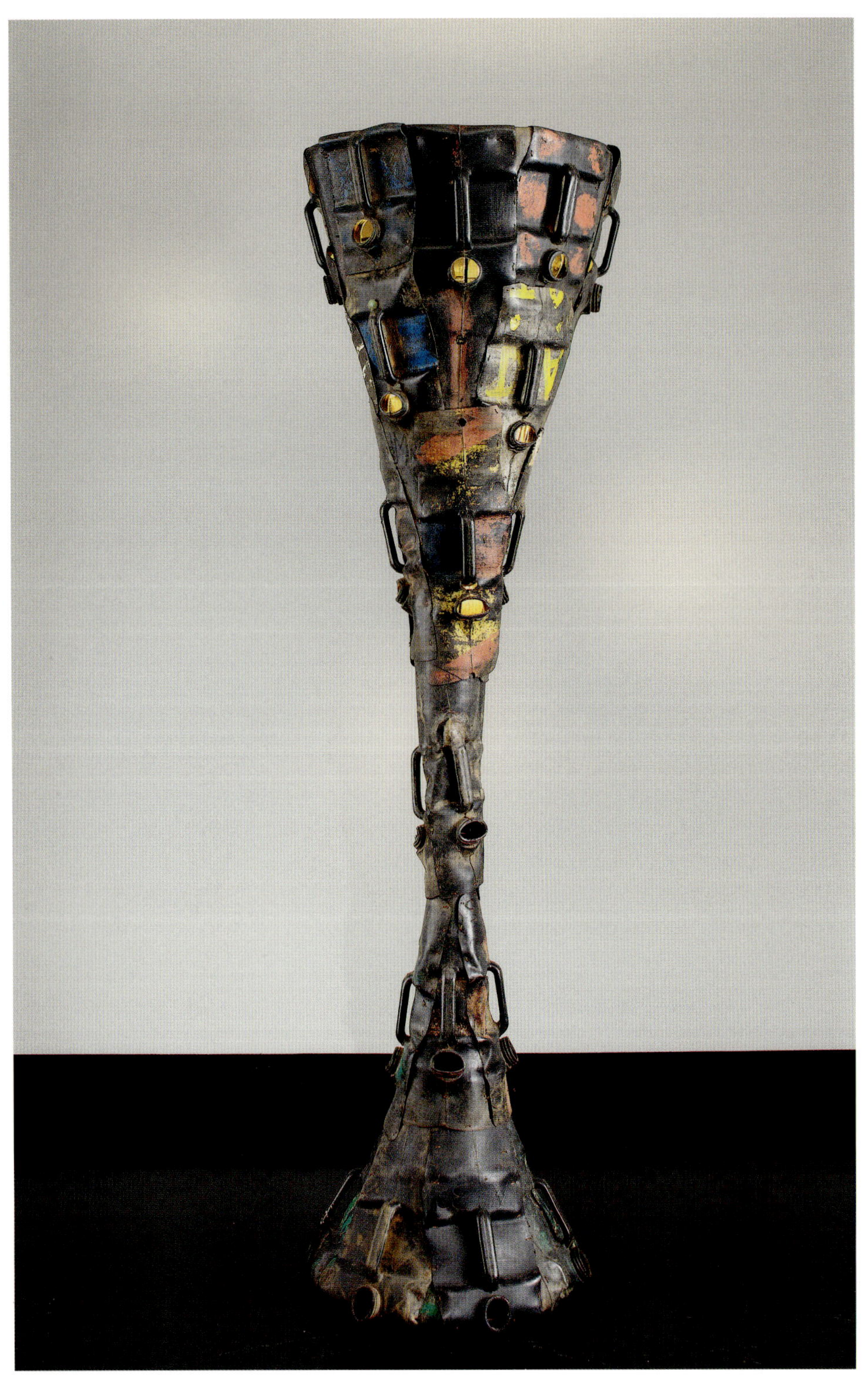

ROMUALD HAZOUMÉ
Tchin-Tchin, BP!, 2010
Found objects, plastic,
and metal
90 × 27 × 27 inches

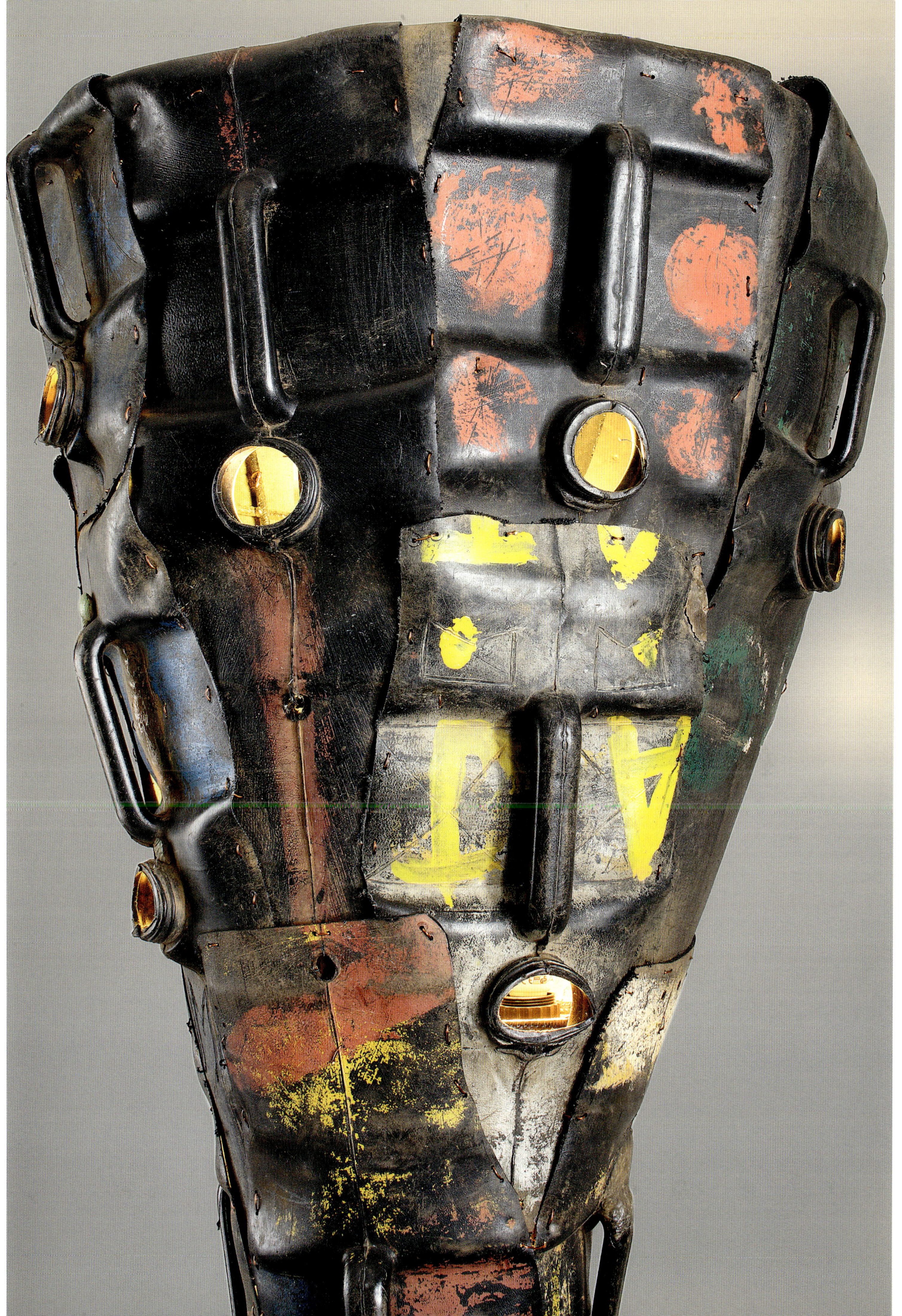

New York's Museum of Arts and Design, which included *Tchin-Tchin, BP!* he notes that the title refers in part to the British Petroleum Deepwater Horizon oil spill off the American Gulf Coast, adding that

Oil disasters are almost a way of life, a truth, for the people living in the Niger Delta, forced to suffer the toxic consequences of oil spills in the fragile wetlands of the region. They receive no actual benefits from the petroleum industry, only suffer the effects of excessive environmental destruction. I have used iconic oil canisters to construct an ironic working-man's version of the crystal champagne flute … a cynical salute to the "oiligarchs"—Your very good health! Tchin-Tchin! (Cheers!)[5]

Reflecting the well-used nature of his found materials, Hazoumé's sculptures typically evoke the fragile and at times damaged feeling so evident in *Tchin-Tchin, BP!* The works communicate this vulnerability far more strongly in person than in photographs, and an undercurrent of pathos and loss mitigates their amusing visual puns. Such masks as *Dan* (1992), with its phone-handset nose, and the jaunty *La Reine* [The Queen] (1995) retain an aura of world-weariness despite their initial charm. All of this work offers a distinct tension between what these things were when they were bright, new products of industry and what they have now become in the artist's hands, namely, testimonies to lives of economic struggle and daily improvisation common across Africa today.

Viyé Diba has long been known as a painter and object maker whose poetic, largely abstract works incorporate a range of found materials and substances drawn from the natural and urban environments of Senegal. Since 2004, following a trip to Rwanda, his work has taken a decidedly political turn, resulting in an outpouring of complex installations tackling such topics as immigration and urban conditions across the continent. In *Nous sommes nombreux, et nos problémes avec …* [We are numerous, and our problems with that…] (2008), Diba fashions a chaotically ordered vision of density in the contemporary African city. His installation incorporates hundreds of components that visitors have to walk over, around, and on top of, recreating for them a visceral sense of the overwhelming layers of visual and commercial activity in such urban centers as Dakar, where Diba lives and works. Specifically, Diba is interested in the "plastification" of the economy and the extent to which the contemporary economy has "packaged" urban citizens as consumers. On one wall of the installation hangs a series of photographs, each of which, according to Diba, illustrates a vexing social or economic problem. A number of the photos

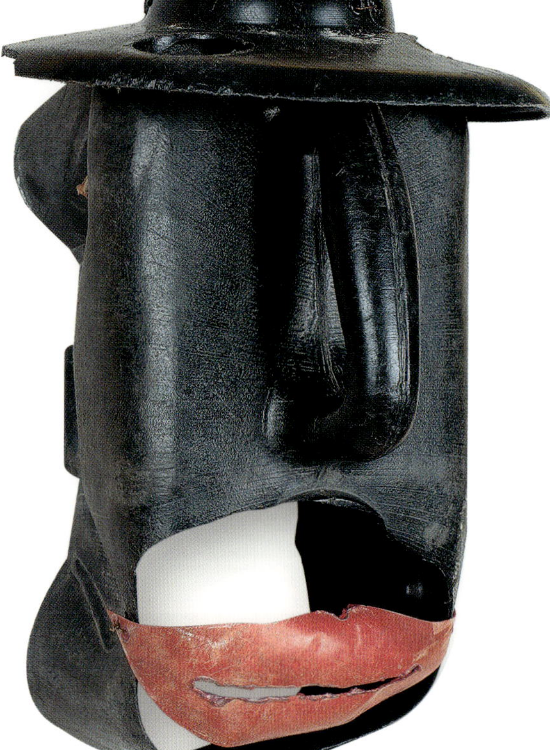

LEFT TO RIGHT

ROMUALD HAZOUMÉ
Le Reine [The Queen],
1995
Found objects, plastic,
metal, and acrylic
9 × 7¼ × 7½ inches

Dan, 1992
Found objects, plastic,
metal, and rubber
15 × 16½ × 6 inches

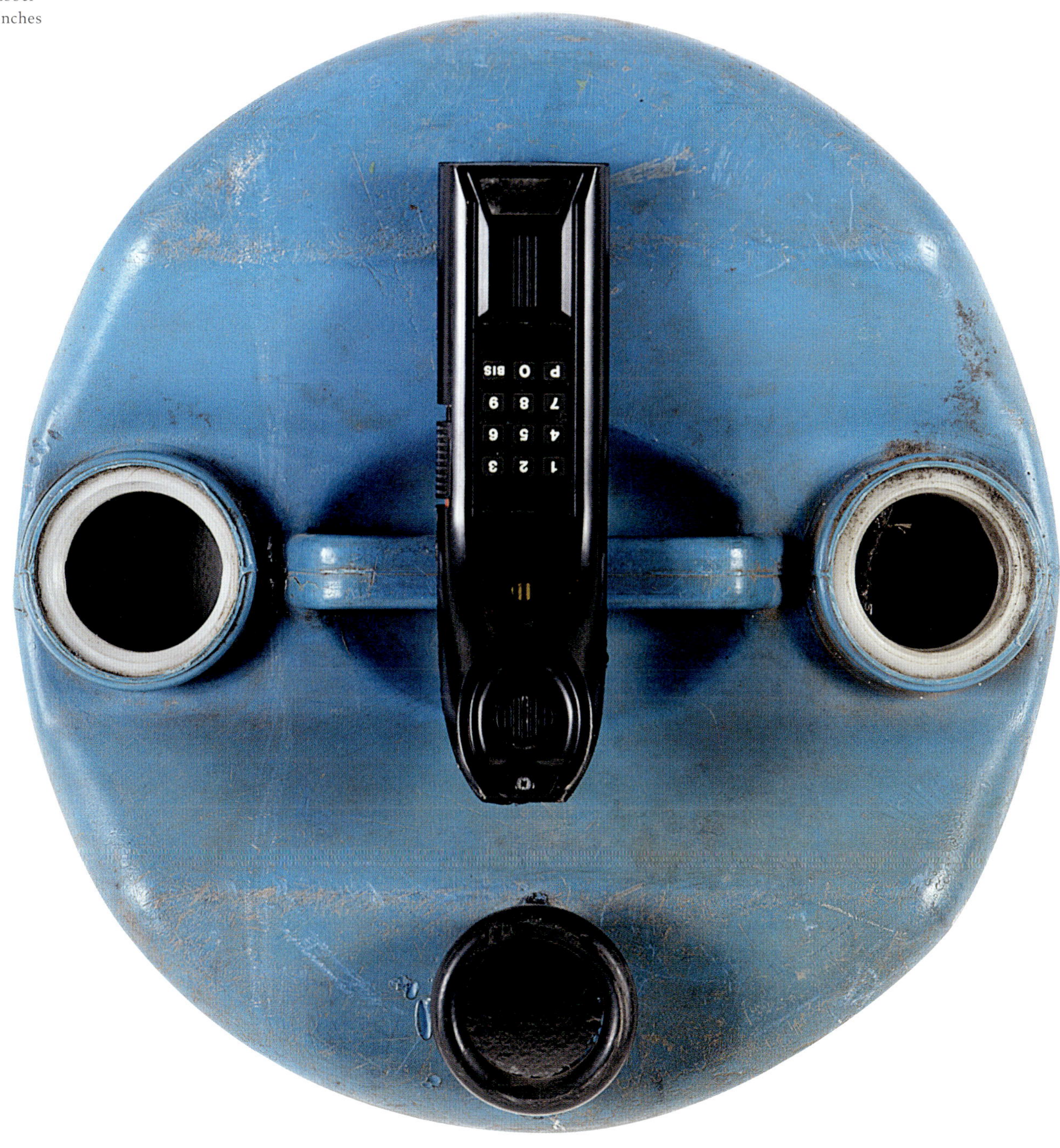

     John S. Weber

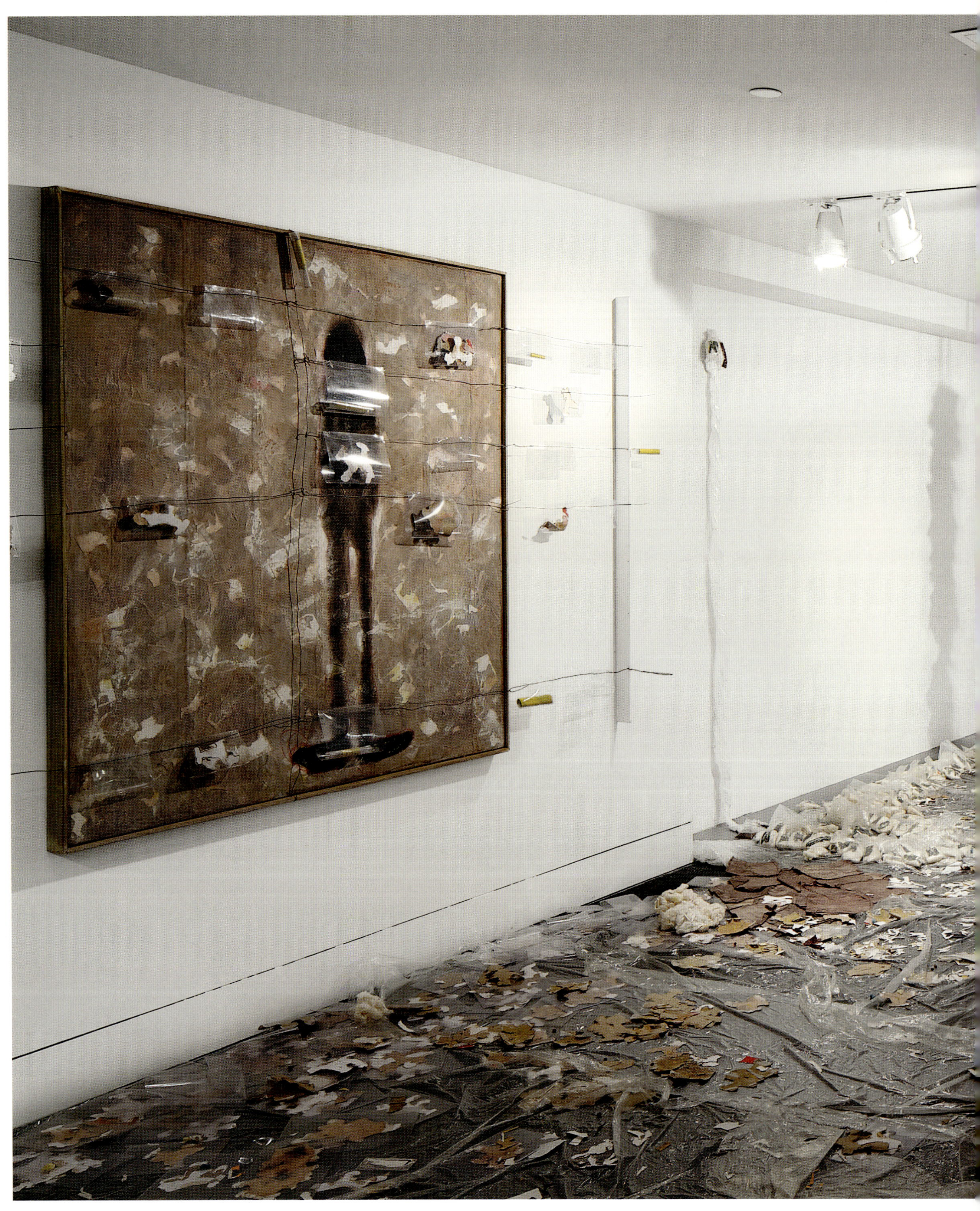

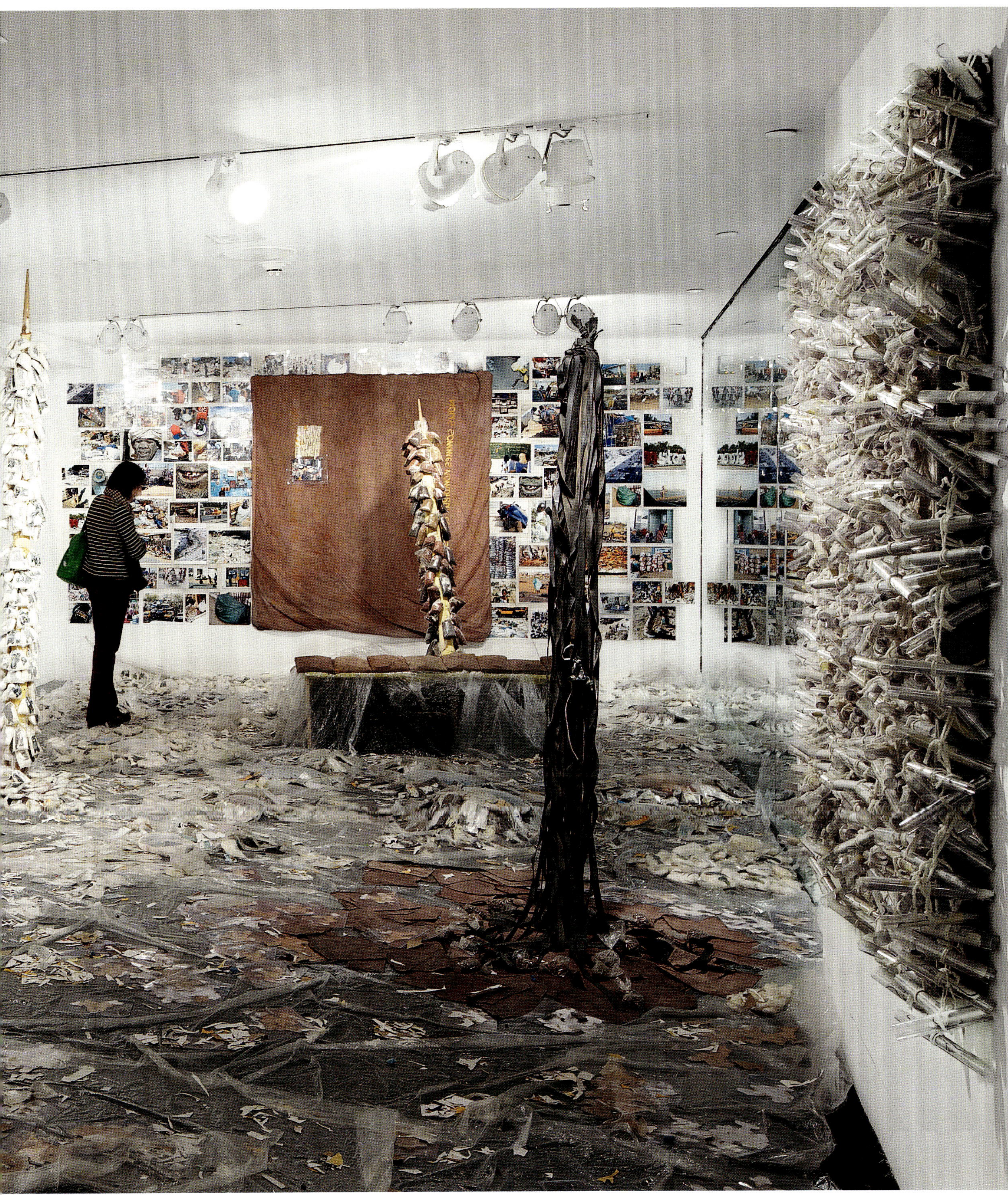

John S. Weber

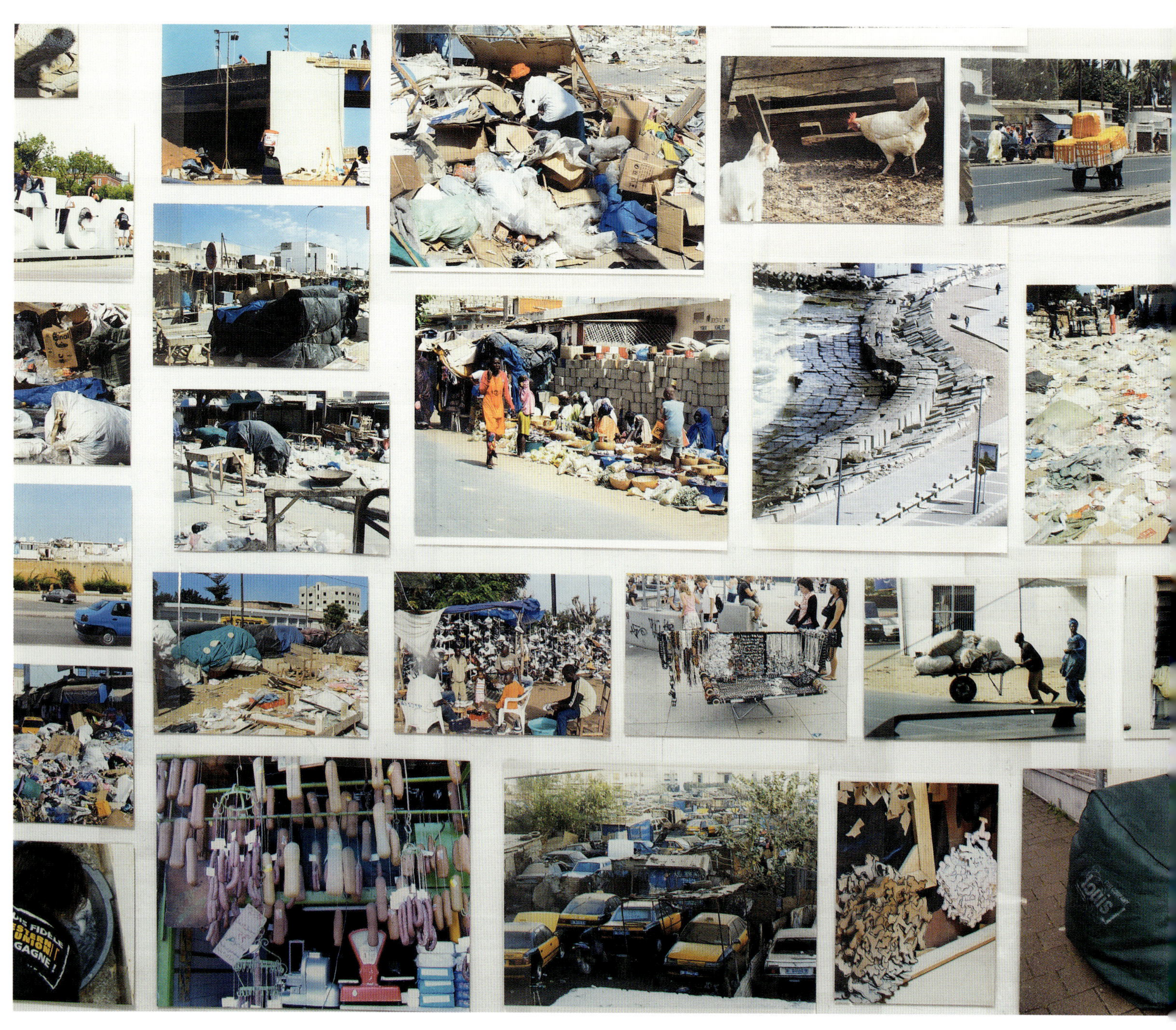

Found, Remade, Repurposed, Transformed

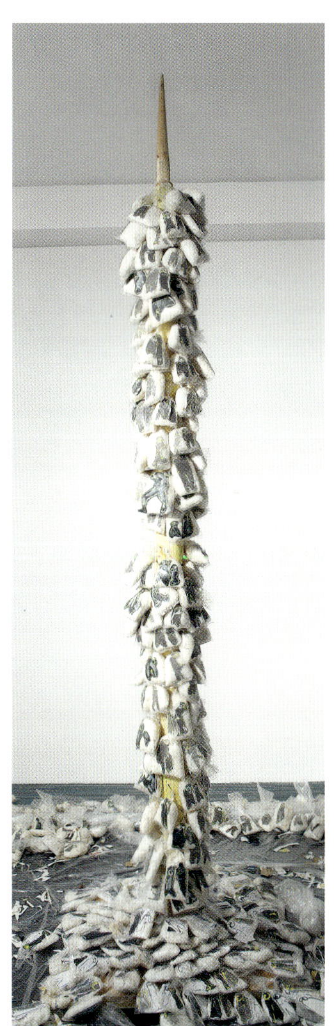

reflect the explosion of the underground or "informal" economy of Dakar and other African cities, where more than half of the population works off the books in improvised businesses and street-corner stalls.

Through a range of visual and compositional strategies, Diba creates an analog for the city in *Nous sommes nombreux*. At the rear of the installation, two carefully gridded and numbered panels map an imagined urban population into a series of squares. For Diba, these numbered panels suggest the rationalized city—the well-ordered, regulated, and duly accounted domain of banks, corporations, and attendant economic structures, all of it either ignored or circumvented by the informal economy. Throughout the installation he has placed plastic bags filled with swatches of raw cotton and crude sketches representing consumers, alluding to the fact that Africa's rich natural resources and raw materials, such as cotton, are routinely sold cheaply to other countries, which then turn them into manufactured products—cloth and clothes—and sell them back to Africans. In a wall piece called *Parchemin de la Consomation* [Parchment of domestic consumption] (2010), hundreds of plastics scrolls contain crude drawings of faucets, directly referring to the metal faucet in Diba's backyard that he says breaks and needs replacement every

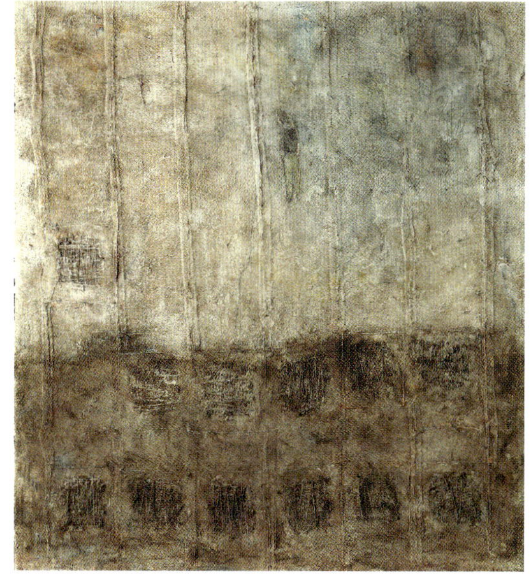

few weeks, a testimony to planned obsolescence and the cheap consumer goods flooding African markets. In Diba's view, both Africa's raw materials and its people are being industrially processed, packaged, and effectively commodified, but ultimately not for the greater benefit of Africans themselves.

The practice of using found materials goes far back in African art, and it has become an accepted aspect of Western modernism as well since Picasso first glued a piece of commercially printed oilcloth depicting chair caning onto a Cubist painting in the early twentieth century. Found objects bring their own voices into the work of art, as well as voices from the world. They anchor the art to a specific time, place, and history. The artists in *Environment and Object* champion this capacity of the found object to embed a geographic, economic, and social context in the work of art. Whether based in Africa, Europe, or the United States, these artists employ found materials to link their work directly to issues of personal, local, and ultimately global concern. At the same time, El Anatsui, Bright Ugochukwu Eke, Nnenna Okore, Romuald Hazoumé, and Viyé Diba all reveal the poetry of commonplace and humble materials, providing visual pleasures that please all the more for their economy of means.

1. Lisa Binder, ed., *El Anatsui, When I Last Wrote to You about Africa* (New York: Museum for African Art, 2010), 18.

2. *Fold, Crumple, Crush: The Art of El Anatsui*, a film by Susan Vogel. Excerpted by the Royal Ontario Museum in conjunction with the exhibition *El Anatsui, When I Last Wrote to You about Africa*, http://blog.rom.on.ca/?p=2181 (accessed December 2010).

3. *Fold, Crumple, Crush* http://blog.rom.on.ca/?p=2181 (accessed December 2010).

4. Gerald Houghton and Sylvester Okwunodu Ogbechie, eds., *Gawu / El Anatsui* (Llandudno, Wales: Oriel Mostyn Gallery, 2003), 21.

5. Romuald Hazoumé, artist statement, *The Global Africa Project*, Museum of Arts and Design, New York, November 17, 2010–May 15, 2011.

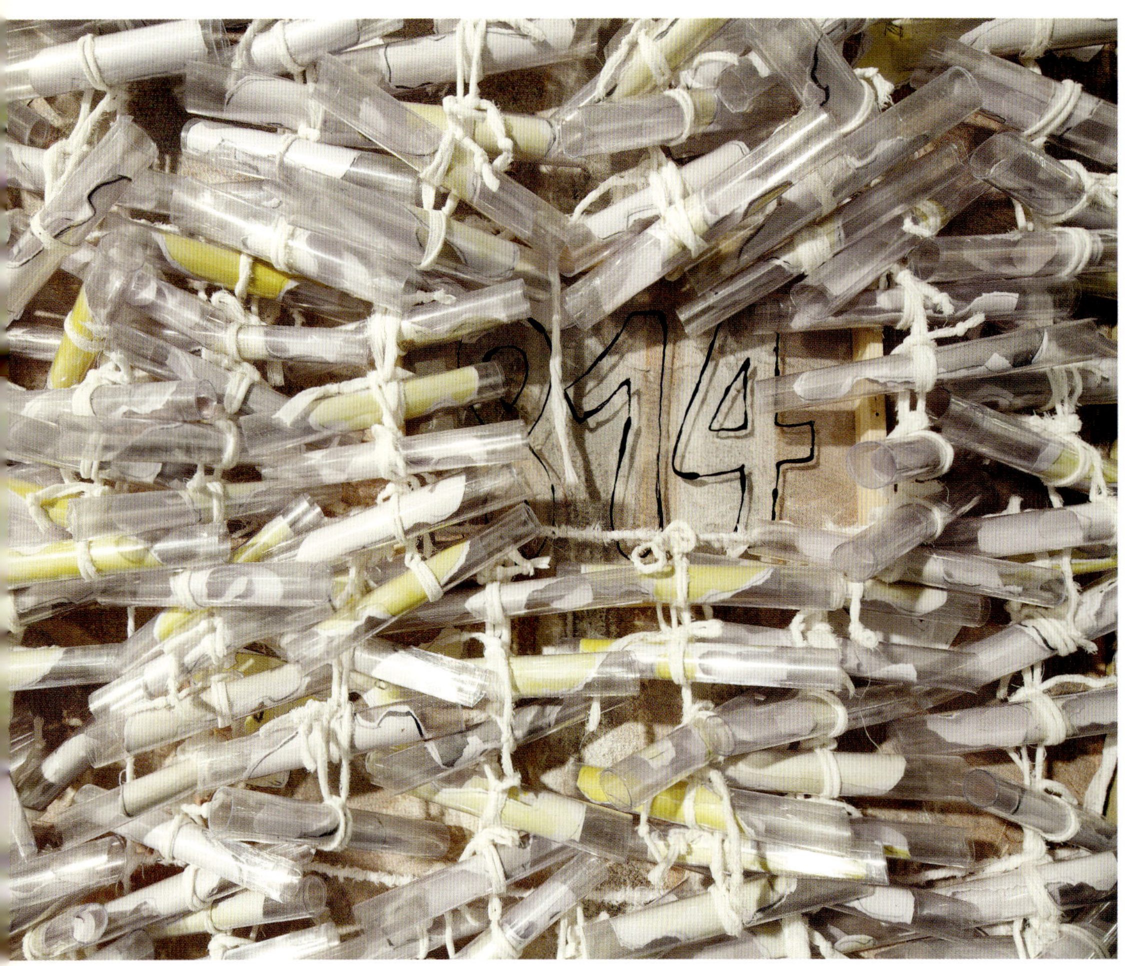

VIYÉ DIBA
Parchemin de la Consommation
[Parchment of domestic consumption], 2008 (detail)
Mixed media
67 × 66¼ × 7⅞ inches

Mark Auslander

"She Speaks with the Wisdom of God"

Traversing Visible and Invisible Worlds in African Environmental Arts

a t the start of the 1988 planting season in eastern Zambia, Mama Zeriya Mwanza, an Ngoni community matriarch and a respected traditional healer, gathered her grandchildren and great-grandchildren to warn them of an approaching drought of unprecedented proportions. Strange behavior and sound patterns by insects and migrating birds, as well as violent interactions in the human world, all foretold dramatic declines in rainfall; hence, selected young people of the lineage needed to travel with her into the bush to learn to gather nutritious roots and fruits and begin to cultivate older drought-resistant cassava strains. "You must do all of this," she said, "if you are to travel the path that lies ahead." Her granddaughter Alice voiced her puzzlement: "What path, grandmother? Can you be clearer?" Zeriya's son turned to Alice and admonished her: "You must learn to listen to her with your heart and soul, to think upon her words by yourself. For when old age speaks like this, she speaks with the wisdom of God."

Taking this incident as my point of departure, this essay explores some intersections among cosmology, indigenous healing, and the apprehension of environmental crisis in contemporary African arts. In diverse African therapeutic traditions, biophysical affliction is broadly understood as symptomatic of broader disruptions in social and spiritual continuums. Illness is properly remedied through the restoration of balance between the visible world of the living and the invisible world of ancestral forces, managed through persons and objects that mediate between visible and invisible domains. This widespread cultural sensibility informs diverse engagements with environmental transformation by contemporary artists across Africa and its diaspora. Images of toxicity, pollution, hyper-consumption, and scarred landscapes hint at underlying social and cosmological imbalances beyond what the eye can normally grasp. The work of the artist, like that of the healer and diviner, opens the viewer to dynamics on the other side of the great boundary zones, thresholds at times evoked by reflective surfaces: the sheen of falling water, the ripples of textiles, even piles of discarded plastic. These practices of seeing anew are at once social and spiritual, diagnostic and therapeutic, polemic and joyous—taking us on dizzying journeys across domains of experience to glimpse simultaneously our current predicament, the arduous work of restoration, and our potential salvation in an expanded, reunified community.

Case Study: A Song about Chemical Fertilizer

Before turning to the extraordinary works on display in *Environment and Object*, it is worth noting that these kinds of aesthetic ventures in Africa are not limited to "high art" works exhibited in museums and galleries. As grandmother Zeriya issued her puzzling warning to her grandchildren, eastern Zambia's Chipata District was abuzz with recently composed popular songs commenting on environmental crises associated with World Bank–backed agrarian policies and increasing dependence on chemical fertilizer. The lyrics of one such song were frequently cited in village conversations, especially by Zeriya and other elderly women:

Feteleza wa mwalunda	Fertilizer, which you have made expensive,
waleta wanga.	has brought [witchcraft] poison.
Wana wanga wanipaye	My children will kill me.
Wana wanga wanipaye	My children will kill me.
Feteleza waleta wanga.	Fertilizer has brought poison.

As it rapidly spread through the southern rural Chipata District, the song inspired excited debate, fiercely opposed interpretations, and prophetic proclamations in the Ngoni communities where I did fieldwork. For a decade, under pressure from the International Monetary Fund and other international institutions, the regime had strongly encouraged rural smallholders to purchase and use chemical fertilizers to grow the hybrid maize that ensured a steady supply of subsidized, industrially milled maize meal to Zambia's urban poor. By the late 1980s, the nation's deepening economic crisis had inflated fertilizer prices and stymied its efficient distribution, threatening rural cultivators with mass debt and serious hunger.

Yet, while the song excited everyone, villagers reached little consensus about what it meant. Understandings of the song varied markedly according to generation and gender. Zeriya and other elderly Ngoni women interpreted it as lamenting their loss of economic security or independence, as chemical fertilizer became inaccessible. Hybrid maize and chemical fertilizer had, by and large, expanded rural women's economic possibilities; "fertilizer" had offered them a chance to support themselves if they were abandoned or their husbands were perennially drunk. They emphasized that under successful conditions, hybrid maize cultivation and chemical fertilizer lowered the amount of time women spent tending crops, freeing them to engage in a range of other money-earning activities, especially after the harvest, when women would otherwise laboriously pound and process local-strain maize stocks.

Fertilizer, many women stated, freed them to earn and spend more money on children's food. Yet the risks ran correspondingly high in those female-headed households committed to hybrid maize; as elsewhere in the region, women producers tended to be denied credit by official agencies. In the months preceding planting, many village women thus brewed beer, took short-term marketing journeys to urban centers in Zambia and Malawi, and engaged in numerous ingenious schemes to raise sufficient cash to purchase bags of fertilizer.

Thus, for many women, this song poetically encapsulated their fear that the rising price and unpredictability of fertilizer would undermine their material and symbolic capacities to feed their children and grandchildren. As one elderly female head of a household explained to me, "My children's cries of hunger will come and I won't stand the pain of them. The cries will just kill me, going straight to my heart. That is why they say fertilizer is like poison. It will just kill me." (In this connection, it is interesting that when I asked several women for the words of this song, they initially mistook the word *wanga* [poison] for the word *njala* [hunger]: "Fertilizer has brought hunger.")

In contrast, older men often glossed the song as a warning against chemical fertilizers altogether. These former labor migrants strove to retain patrilineal and patriarchal control over multigenerational households by monopolizing bride-wealth cattle transactions—the basic act of male social reproduction. For older men, the song expressed the danger that their sons—greedy for chemical fertilizer—would "kill" them through *wanga* poison and then sell off their cattle for cash to purchase fertilizer. Chemical fertilizer, after all, is purchased by money and generates money. Its use contributes to the growing individuation of production and the hiving off of small nuclear family households. Young men thus continually berated their elders to sell off extra cattle for fertilizer; but older men nearly always refused, insisting that while the money realized through fertilizer was ephemeral and "slipped through people's fingers," cattle constituted "true wealth" that "lasted forever."

The ChiNsenga and ChiBemba term *wanga* is usually translated as "poison" or "witch-charm," but more precisely, it denotes an extraordinary innate capacity to influence and work upon the outside world, either for negative or positive ends. It may be embodied in the actor him- or herself or concretized in a material object. As a substance, *wanga* is said to be placed by evildoers in men's beer at beer parties; nowadays, it is usually held to be some sort of industrially produced toxic substance—mainly battery acid, pesticide, herbicides, or chemical fertilizer—possibly combined with various indigenous toxic barks or roots.

The song thus partly warned young men not to engage in parricide. But it also decried a pervasive change in the very fabric of existence. Like other *wanga*, fertilizer is a substance of extraordinary positive potential but one that under current conditions causes manifold afflictions. For all its power, chemical fertilizer intercepts and alters a process of biophysical transformation associated with food. Eating food grown with fertilizer fails to nourish children, insisted many older males, and instead stunts their growth and renders them more susceptible to malaria. What is more, elder men insisted, fertilizer leads to the continual leakage of value and wealth from the body politic, poisoning the social hierarchy, even as it poisons the land itself. At beer drinks, old men berated me because in the old days the soil was productive without chemical fertilizer, "but now that you white people have brought us fertilizer, the land cannot do without it."

In contrast young men directed me to other songs about chemical fertilizer. Most beloved was this evocative song, in the *kanindo (kalindula)* style often heard on regional radio:

Tiyene	Let's go
ku munda	to the field.
feteleza ndilibe	I don't have fertilizer
mwandisautsa	You have made me suffer
Abale tiyene	Friends [brothers] let's go
tikalime	to cultivate
feteleza ndilibe	I don't have fertilizer
mwandisautsa	You have made me suffer

Like the "poison" song, this song evoked commentaries about both national institutions and the internal dynamics of local kinship and residential units. Many younger informants gloss the elided "mu" (the second person plural/formal) in "mwandisautsa" (you have impoverished me) not only as the state but also as the elder men of the village, who are always berating young men to be industrious and go out to the fields, even when there is no fertilizer and the effort is doomed to futility.

Young men often told me how much they usually enjoyed the morning walk out to their fields and gardens, where their fathers and uncles could not directly observe or supervise them. Hence, the extra pathos of the song, which starts out so optimistically: just at the moment when a youth ought to feel most liberated from external control, he is brought up short, painfully reminded of his continued dependence on older men and on the distant state. In this sense, the song evoked the everyday landscape experienced by most young men, with the lack of chemical fertilizer tangibly and oppressively evident in their wilted maize fields and shrinking economic horizons.

CLOCKWISE FROM TOP LEFT
ROMUALD HAZOUMÉ
Anna Fiata, 2002
Found objects, plastic,
metal, and synthetic hair
18 × 12½ × 6½ inches

Brut, 2001
Found objects, metal,
plastic, and rubber
13½ × 14¾ × 3 inches

Patatipatata, 2002
Found objects, plastic,
metal, and synthetic hair
14½ × 8 × 6½ inches

Gris-Gris, 2004
Found objects, plastic,
and synthetic hair
14½ × 10⅞ × 7 inches

Kaleta, 1992
Found objects,
plastic, seeds, cowries,
rubber, and metal
21 × 14 × 7½ inches

Claudia Maigre, 2005
Found objects, plastic,
and synthetic
and cotton thread
11½ × 10½ × 8 inches

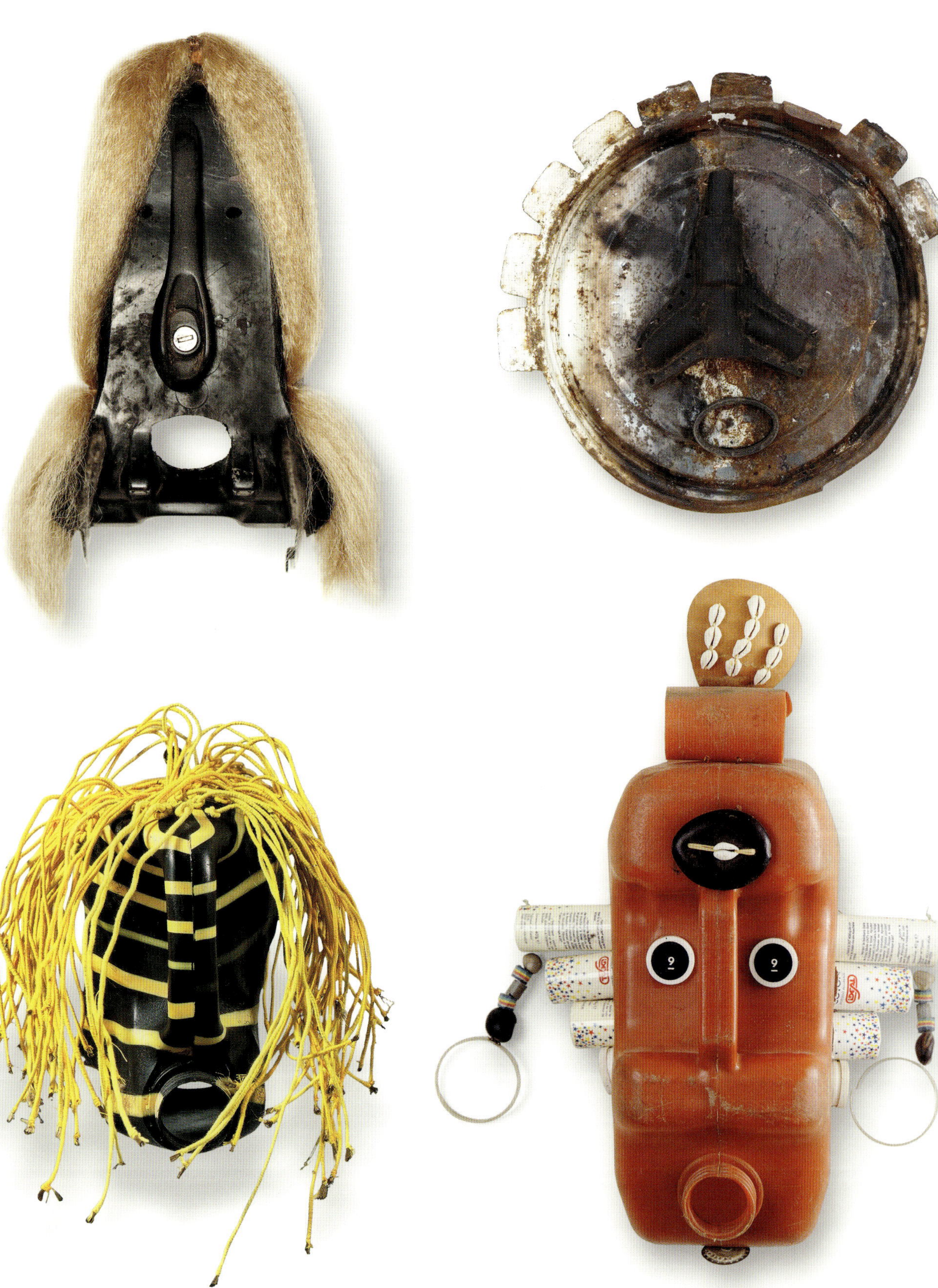

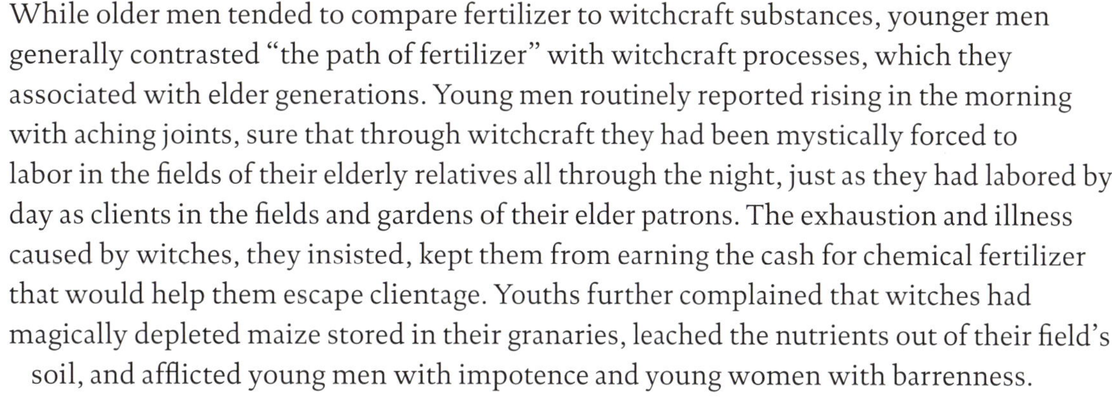

While older men tended to compare fertilizer to witchcraft substances, younger men generally contrasted "the path of fertilizer" with witchcraft processes, which they associated with elder generations. Young men routinely reported rising in the morning with aching joints, sure that through witchcraft they had been mystically forced to labor in the fields of their elderly relatives all through the night, just as they had labored by day as clients in the fields and gardens of their elder patrons. The exhaustion and illness caused by witches, they insisted, kept them from earning the cash for chemical fertilizer that would help them escape clientage. Youths further complained that witches had magically depleted maize stored in their granaries, leached the nutrients out of their field's soil, and afflicted young men with impotence and young women with barrenness.

While fertilizer offers men the capacity to produce value and ultimately to reproduce themselves, it is nonetheless capricious, continually escaping male encompassing and control. For many Ngoni young men, fertilizer is rather like a lost love—at first seeming so wonderful and full of promise but now turning out to be not all she seemed. Hence, the appropriateness of the *kanindo* guitar genre, so often used in ballads of lost love.

Environment and Object

It is helpful to keep in mind these songs, as well as the striking variations in their interpretation, as we consider contemporary African artistic production oriented toward environmental crisis. Like the composers and performers of the songs, many visual artists sense that visible environmental challenges are symptomatic of larger spiritual and moral ruptures in the body politic. A visible tear in the earth often evokes a less visible wound in the fundamental fabric of the social world, in a way that encompasses ancient spiritual traditions and modern political and economic preoccupations. These powerful works, however, do not always express a fully formulated ideological position on the causes and implications of environmental distress; rather, they speak to audiences on multiple manifest and latent levels, generating a proliferation of possible interpretations.

Consider, for example, the art of **Romuald Hazoumé**. Incorporating the plastic jerry cans normally used for transporting gasoline, his pieces recall the shapes of sacred masks. In tradition-based masquerade ceremonies in West and Central Africa, the mask is a tangible evocation of a revered ancestor or spiritual figure, whom the human dancer embodies in the course of dance performance. Made of wood from trees that are often understood as the homes of the honored dead, the mask is usually a switch-point between this world and its opaque counterpart;

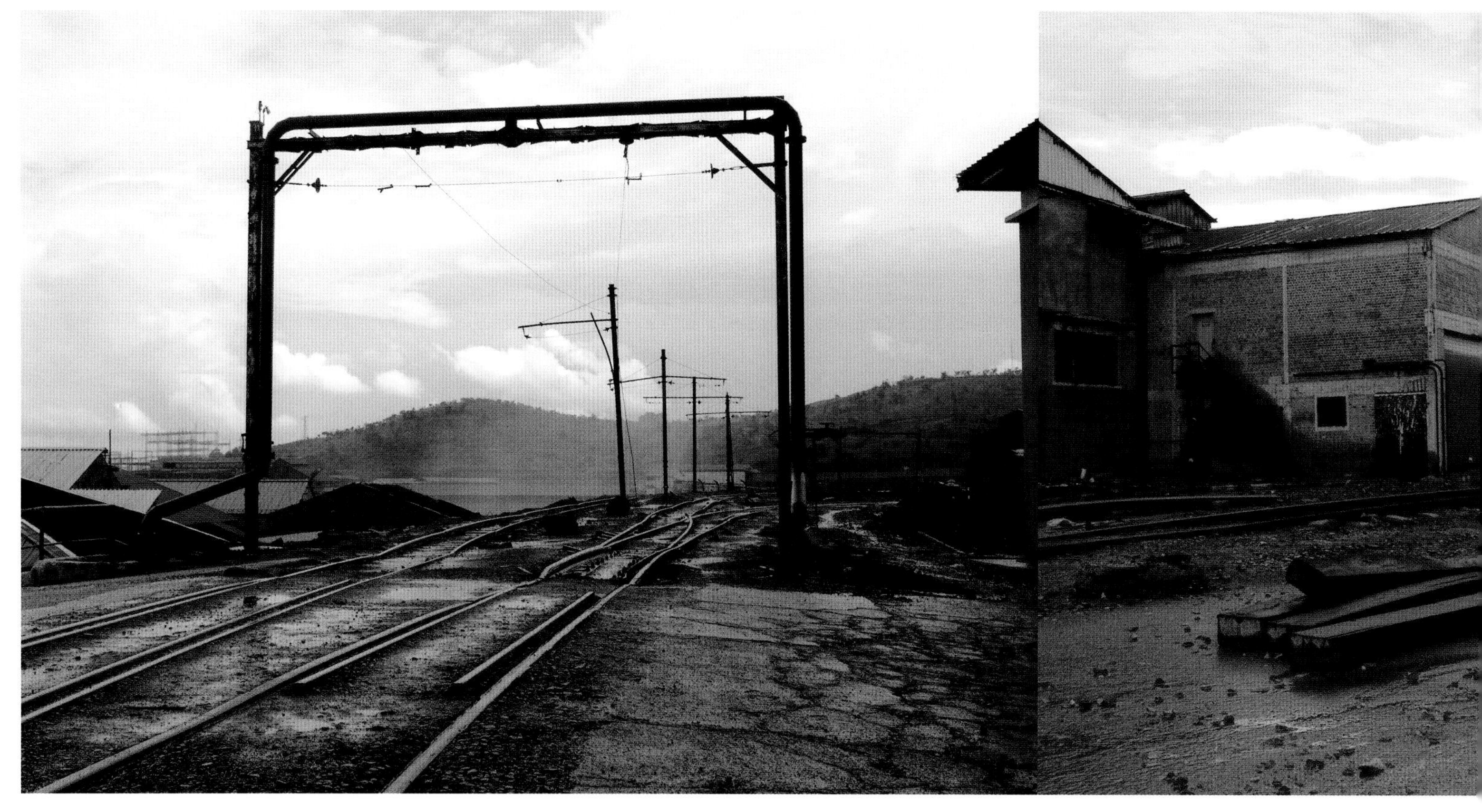

     "She Speaks with the Wisdom of God"

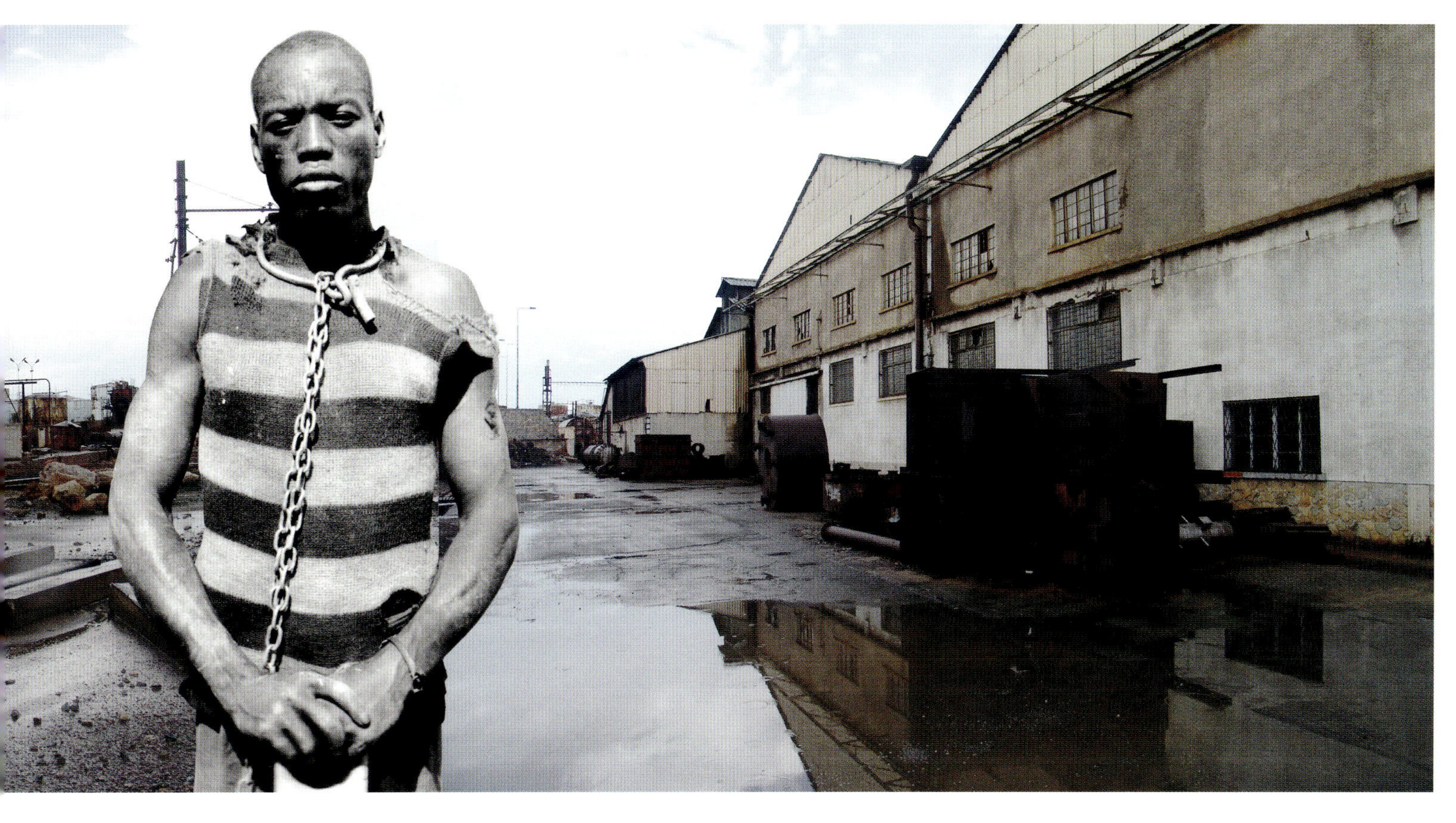

the whistles and enigmatic sounds of the spirit world are often projected through the mouth of a mask, a channel through which life-giving forces enter into the human domain. For better or worse, oil is the lifeblood of contemporary African economies; how appropriate, then, that in Hazoumé's works, our eyes quickly go to the mouth-like spouts, from which flow this potent substance to power vehicles of all kinds. Is Hazoumé celebrating the magic of petroleum or decrying the ways in which oil has distorted African economies and moralities? As with the fertilizer songs, any single interpretation would be overly limiting, yet all can take joy in the artist's playful integration of the spiritual and the material, the old and the new, the venerable and the mundane.

Recycled materials also figure in the translucent walls constructed by **Bright Ugochukwu Eke**, often out of discarded plastic water bottles. Although the works can be read as decrying the waste and violations of a throwaway society, they also evoke the transformative and life-giving power of water itself. In diverse African ritual processes, water functions as a universal solvent, breaking down conventional categorical oppositions (life/death, male/female, young/old) in order to allow for creative, productive recombinations. Such may be the case in Eke's labyrinthine, liquid-like installations, which, like West African–inspired bottle trees, promise us glimpses of worlds beyond our normal capacity to see. Similarly, **Nnenna Okore**'s assemblages incorporate discarded refuse to create marvelous complexes that evoke the full range of visual experience in Nigerian society, from piles of market produce to towering spirit shrines.

The refuse of industrial civilization is also brilliantly recycled in the works of **El Anatsui**, who transmutes everything from half-filled liquor bottles to aluminum bottle caps into shimmering, cape-like assemblages that evoke the Ewe *kente* cloth of his homeland. West African cloth designs often have embedded within them enigmatic messages accessible only to the initiated; these forms of worn esoteric knowledge bind together lineage members in a fabric of social and spiritual continuity that comprises ancestors, the living, and those yet to come. El Anatsui's glorious pieces remind us of the age-old African insight that apparent loss may be subsumed in cycles of productive regeneration. His intriguing title, *Some Still Come Back* (2005), signals the creative recycling of discarded aluminum caps but might also be read as evoking the returning potency of the dead who, while once lost, continue to infuse the living world with their vital, mysterious energies.

In a comparable vein, Congolese artist **Sammy Baloji** moves back and forth across temporal, spatial, and spiritual domains. His digital images depict the violated

GEORGIA PAPAGEORGE
Genesis, Crucifixion, Aftermath (from Africa Rifting), 2003
Ultrachrome prints
3 parts, each 69⅜ × 47¼
Installation view, *Environment and Object · Recent African Art,* Tang Museum, 2011

Africa Rifting: Lines of Fire Namibia/Brazil, 2001
Looped DVD, 15:39 minutes

   "She Speaks with the Wisdom of God"

landscapes of Central Africa, from which the mineral abundance underlining colonial wealth has been extracted for over a century, and the traumatized bodies of the workers who have extracted it. Although set in ravaged postindustrial ecoscapes, his work recalls the careful and disciplined choreography of Central African masquerades, in which mythic figures are summoned forth to impart wisdom to their posterity. Thus, even when one of his 2006 photographs depicts a pith helmet–clad Belgian official from the past, viewers from Katanga province and environs may understand a present-day denunciation of postcolonial warlords and transnational corporations that continue to profit from extractive predation. Similarly, the right-hand panel of *Untitled 13* (2006) centers on a male figure situated in front of industrial warehouses, wearing a tattered garment of black-and-white stripes, as a metal chain hangs from his neck. The left-hand image shows a barren railroad scene. The conjoined images move us both back and forth in time and space, recalling the horrors of the Atlantic slave trade, the terrors of a chain gang in the southern United States, the ravaging of the African continent by industrial penetration, and present-day captive labor in the Congo basin. In the words of William Faulkner, "The past isn't dead. It isn't even past."

The steel sculptures of **Sokari Douglas Camp** also powerfully integrate masquerade and environmental symbolism. Her works evoke the strife-torn Niger Delta, where corrupt Nigerian political and military figures have long partnered with transnational oil companies to plunder the region's mineral resources to the great detriment of its people. Drawing on the Kalabari masquerade traditions in which she was raised, Douglas Camp's figures exude the energies of spirit-possessed dancers caught up in a dangerous choreography of love and loss, life and death. Her enigmatic *Teasing Suicide* (2004) shows us a woman, spattered with blood or oil, bearing an AK-47 with its muzzle inside her

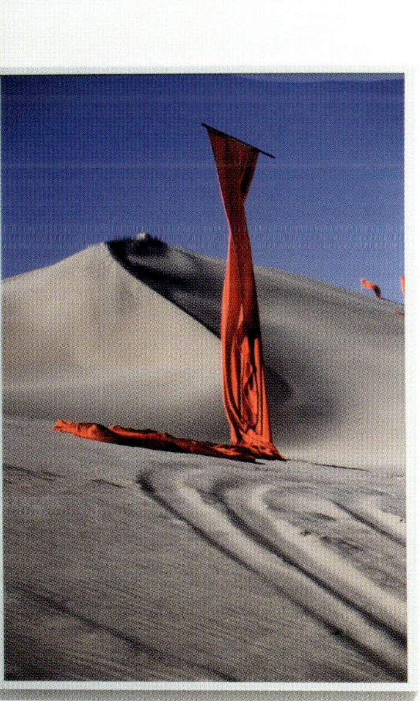
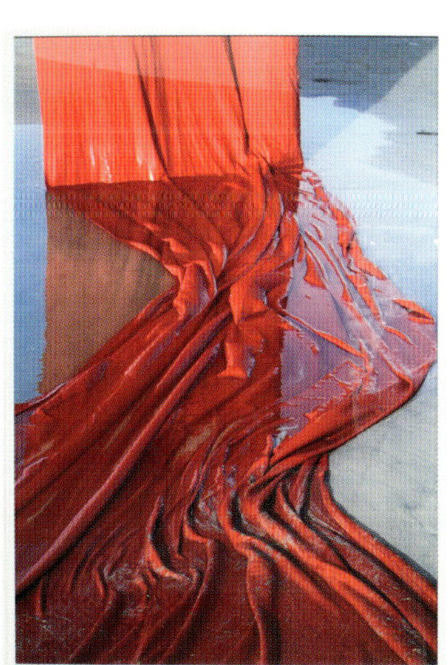
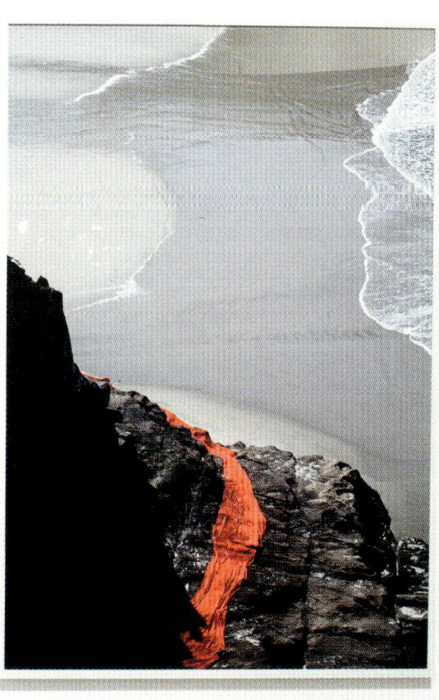
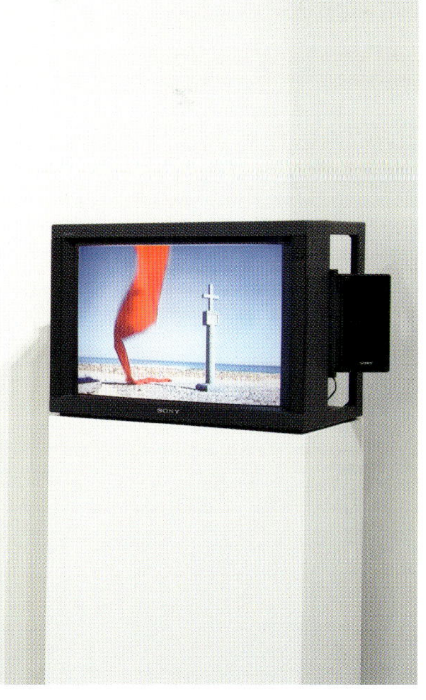

mouth. Are we to understand the figure as having died at her own hands, perhaps to draw attention to her people's plight? Alternately, as with so many other torture victims, should we understand that the death of this resistance figure in detention was cynically labeled "suicide" by the regime? Yet again, the work provides no easy answers but poses vital interrogatives aimed at stimulating critical reflection.

The varied works in *Environment and Object* often make it difficult to distinguish cries of pain from exclamations of hope. In many respects, these projects exemplify the extraordinary power of ritual, and of art, to move people and communities from trauma toward mourning and, finally, back toward the regular flow of life. Consider, for example, **Georgia Papageorge**'s mesmerizing video documenting her performance piece, *Africa Rifting: Lines of Fire Namibia/Brazil* (2001). Not easy to interpret, the piece seems to move from a specific lament for the artist's dead child and the horrors of the Middle Passage toward a cathartic resuturing of the long-divided lands of Africa and Brazil, now linked by artery-like red banners that imaginatively crisscross the Atlantic and honor the oceanic *orisha* Yemayá. The video, like so many other works in the exhibition, calls for repeated viewings by visitors, asking each of us to puzzle through its captivating, ambivalent imagery.

Priests, it has long been observed, "speak for Man to God," while prophets "speak for God to Man." Invariably, prophetic discourse is enigmatic, ambiguous, and painfully difficult for mortals to disentangle—surely how it should be. The labor of interpretation, of pondering mystery in the world around us, constitutes a fundamental form of veneration, a disciplined honoring of the great forces in and beyond human experience. As James Fernandez notes in *Bwiti: An Ethnography of the Religious Imagination in Africa* (Princeton University Press, 1982), profound spiritual revelation in the West Central African religious movement of Bwiti proceeds not through direct lecture but rather by means of "edification through puzzlement." Those who "speak with the wisdom of God," whether they are grandmother Zeriya, a singer about fertilizer in a Zambian village, or a contemporary African artist in a London studio, rarely provide simple answers or easy-to-decipher blueprints. Yet by perplexing us and making us think for ourselves, they inspire us to see with new eyes, to listen with new ears, to become aware of the constant, life-sustaining interplay between the living and those who came before us, between the visible and the invisible. In so doing, they help us fashion for ourselves the tools, the resilience, and the imagination to chart our way along the uncertain paths that lie ahead.

Christopher A. Whann

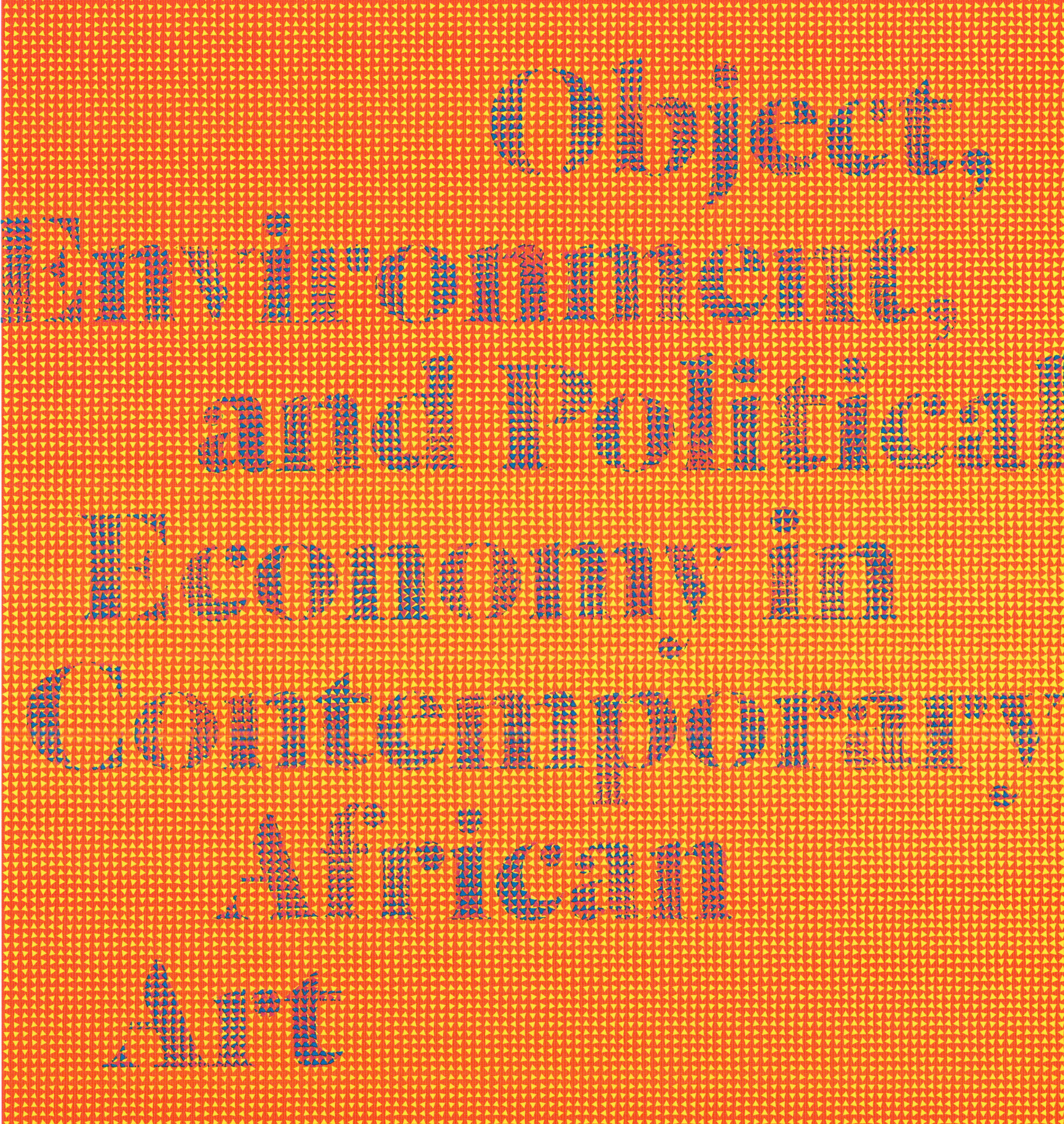

Object, Environment, and Political Economy in Contemporary African Art

fricans and scholars of Africa have long recognized that humans' relationship with nature is a driving force and an essential condition for understanding daily life. African political economy reflects the dialectical manner by which Africans shape their environment and African lives are shaped by it. This exhibit offers visual and visible evidence that this is so.

John Iliffe, the noted historian of Africa, has observed that two of the key themes of African history are "the achievement of human coexistence with nature" and the "defence against aggression from more favoured regions."[1] These themes resonate for students and creators of material culture and for shapers of the African environment. Most people outside Africa think about the continent in connection with civil wars, the socioeconomic catastrophe of HIV/AIDS, the problematic importation of conflict diamonds from Sierra Leone, genocide in Rwanda, and the pirate attacks based in Somalia. Yet Africa is a continent of breathtaking diversity, and many success stories match the media's portrayals of failure.

It is easy to break African political-economic history into several grand eras: precolonial, colonial, and contemporary independent Africa. It is equally easy to look at African regions as somewhat self-contained in terms of their economic dynamics and their histories. Each of these categories can be further subdivided, and each possesses its own fascinating internal dynamics and trajectories.

States, Societies, and Globalization

Political systems in Africa are as old as anywhere else, but they bear little resemblance to the contemporary political landscape—no more than classical Athens and Sparta look like the modern U. S. Congress. Jeffrey Herbst describes earlier African political entities as functions of central control over their hinterlands, with very fluid "borders." In a continent that always has been lightly populated in relation to its geographical size, leaders have been challenged to bring authority and order to these peripheries. Empires like Mali and Ghana rose and collapsed over the years. North African rulers were tied both to the Mediterranean world (as visitors to Carthage can attest), and East African Swahili Coast traders were linked to Southwest and South Asia by the monsoon winds carrying goods and people. Borders and regional divides blurred, insofar as southern African kingdoms of the 1800s migrated as far north as modern Tanzania, and Muslim rule spread well into west-central Africa and along the Swahili Coast from its roots in the Arabian Peninsula and North Africa.

The colonial period brought significant change to that early sort of politics. European state managers, through their imperial offices, claimed domain over regions they had explored and ruled for centuries. Portuguese explorers had maneuvered around the African coastlines since the fifteenth century, circumnavigating the Cape of Good Hope in 1497. Most early European contacts with Africa and Africans were coastal trading stations. Africa more or less provided a stepping-stone to trade with Asia. One need only look at a map of southern Africa, with place names like Alagoa and Delagoa, to see that the important Portuguese-controlled Indian territory of Goa was the true destination for Europeans seeking riches through global trade.

Europeans established chartered companies (proto-corporations set up for risky international ventures) in an effort to make money through these trading centers in Africa. As many scholars have noted, African economies were globalized from the early days. The first truly large-scale commercial operation out of Africa, monstrous though it was, has to be the slave trade. Not only were almost twenty million Africans sold into bondage through the trans-atlantic slave trade, but many millions more were sold through the East African slave trade to the Islamic world; the slave trade shaped not only European states but African ones too. For example, the trade along the West African coast caused small forest states to expand into kingdoms and empires like Ouidah (often spelled Whydah) and Ashanti.

As politics in Europe became ever more complicated in the 1860s and 1870s, exemplified by Germany's unification and the restructuring of great power relations, European monarchs and governments viewed Africa as a place to divert conflicts out of the European core. European explorers traveled through interior parts of Africa, about which they often knew little or nothing. Henry Morton Stanley's "search" for the missionary David Livingstone is the most famous of these journeys; Stanley undertook some of his exploration under the auspices of Belgium's King Leopold II, who sought a personal colonial possession from which to extract wealth, which eventually came in the form of ivory and "red rubber"—red because of the African blood spent to make Leopold rich.

The Berlin Conference of 1884–85 was the result of that European effort to keep the great and lesser European powers happy, as well as Leopold himself. The scramble for Africa and the carving up of the continent followed clear ground rules after the conference; European powers "awarded" chunks of Africa were obliged to "pacify" the inhabitants, establish functioning colonial governments, and ensure "effective

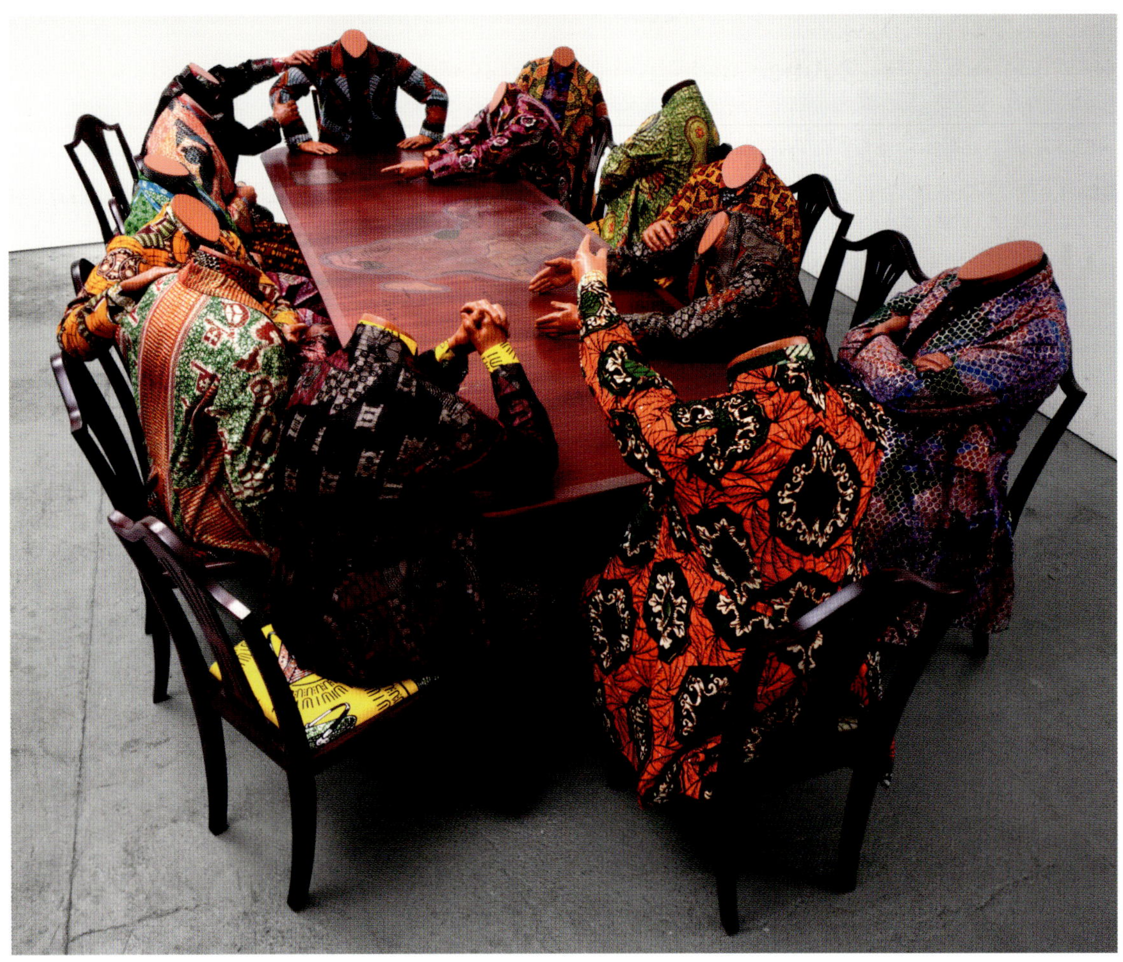

occupation" of their new colonies. Many Africans resisted at great cost. The Germans annihilated Hereros in South West Africa (now Namibia) and killed many more East Africans in the Maji-Maji Rebellion, harbingers of the later Holocaust in Central Europe. Portuguese colonialists killed thousands during the Bailundo wars through the 1910s.

Ghana's 1957 independence began a wave of freedom for sub-Saharan African states, with most African countries achieving independence in the early 1960s and the Portuguese colonies only becoming independent in the mid-1970s. The Spanish granted Western Sahara independence in 1976, though Morocco occupied the region immediately and contests it even now. Zimbabwe (1980) and Namibia (1990) were the last to gain independence from their minority white-dominated governments. The vast majority of Africans living today were born after 1960; so most Africans do not even remember the days when Europeans ruled them.

Natural Resources:
From the Congo Rubber and Ivory Boom to Modern Mining

As the exhibit depicts, African natural resources include both "human" and "non-human." The slave trade made humans into a commercial product. Deep mining, strip mining, and oil drilling number among African countries' primary economic activities in the modern global economy.

The fifty-four independent African countries provide the global economy with oil, diamonds, minerals, and other natural resources. They export cash crops like cocoa, coffee, and tea. They import new and sometimes used goods from Europe, Asia, and North America. While colonialism undoubtedly shaped patterns of global trade and domestic political arrangements, Africans today control their own destinies rather than suffer as objects of age-old political dispensations.

While many African countries possessed road, rail, and shipping infrastructures at independence (testaments to the extraverted nature of the economies), the new African leaders and their public administrators were left on their own to maintain them. State managers determined the future direction for economic development. Many African leaders enthusiastically endorsed Soviet-inspired, state-driven, large-scale, crash industrialization and agricultural efficiency programs. By the mid-1970s, African states poured billions of dollars of domestic resources and foreign aid into failed development schemes. For example, Ghana's leaders invested in the Volta River development project, meant to provide hydroelectric power and a successful aluminum smelting facility. Even before the project could be completed and achieve any of its aims, President Kwame Nkrumah was overthrown, and the military regimes that succeeded him stole millions of dollars, drastically damaged the thriving cocoa-export sector, and drove Ghana into a decades-long economic tailspin. Tanzania's Julius Nyerere, a very popular leader within his country and widely admired throughout the world, implemented a system of collectivized farms called *ujamaa* villages that did nothing to help increase production of staple or export crops. Only by the 1990s have these countries begun the long road back to economic growth and improvements in the standard of living for their citizens.

The wasteful diversions created by failed development projects meant fewer resources for the ongoing battles against nature. Malaria continues as a major public-health challenge across the humid zones of the continent, despite some significant contributions in the last decade from large charitable organizations and governments in advanced industrial countries.

 Christopher A. Whann

Other countries, like the Democratic Republic of the Congo, have never recovered from the disastrous Belgian colonial period, the civil war of the 1960s, the depredations of the Zairian Mobutu era, and the ongoing civil war. And yet none of our cell phones would work without coltan, the metallic mineral extracted by Congolese miners working under extremely difficult conditions in war zones.

Coltan is just one example of the wealth African countries possess. Though the transatlantic slave trade has been illegal for two centuries, Africa provides the world with other remarkable natural resources. In agriculture, the Ivory Coast and Ghana are two of the world's largest cocoa exporters. Kenya and Tanzania export very high-quality coffee and tea. South African wines rate among the best in the world. Southern and central African countries produce such minerals as gold, chrome, copper, cobalt, platinum and titanium for economic growth and industrial production. Botswana produces some of the world's best gem-grade diamonds, with a government ranked among the best in Africa.

As several pictures in the exhibit represent, crude oil and petroleum derivatives have reshaped the African landscape. Of the United States' fifteen largest oil suppliers in May 2010, four are African (Nigeria, Angola, Algeria, and the Republic of Congo [Brazzaville]). Nigeria alone exports over a million barrels of oil a day to the United States. The environmental and human impacts on Africa are staggering. Some scholars have hypothesized about Africa's "resource curse" to explain why African countries have not benefited as much from oil as imagined or hoped. Those who fight against governments that destroy communities while wasting such a national patrimony are often imprisoned or tortured or, like the Nigerian martyr Ken Saro-Wiwa, executed (see page 53).

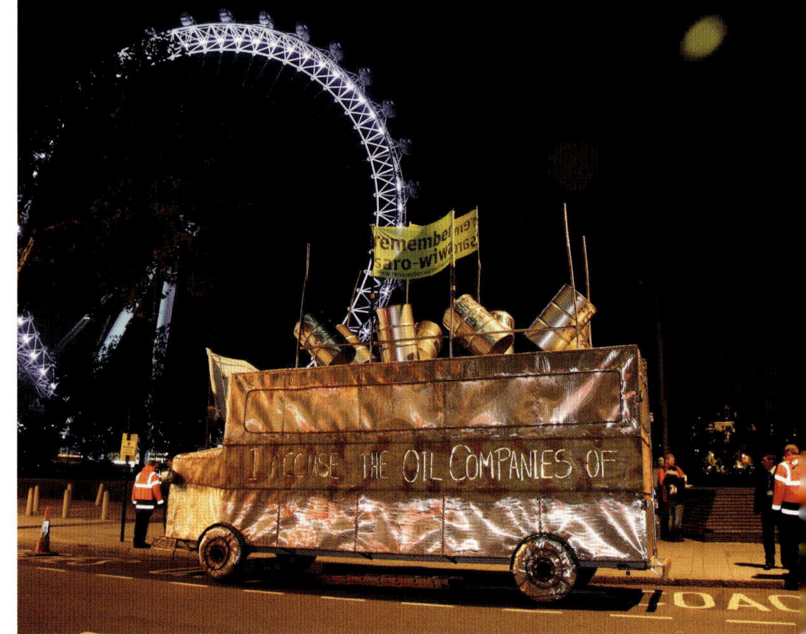

Signs of Hope?

Though African countries' ties with former colonial overlords remain strong, the Chinese government and Chinese business are the newest investors. Rapid growth in the Chinese economy keeps pushing up demand for oil and oil-based products. The faster the Chinese economy grows, the more that nation needs African natural resources and markets for its exports. This trade relationship may be a mixed blessing, but one could argue that it is a way for Africans to diversify their own markets and expand the pool of interested investors.

In the 1990s and 2000s, governance improved across the continent. Military rulers stepped down by the dozens; civilian regimes and multiparty systems arose, and more societies opened to political debate. The dead hand of intrusive but ineffective management permitted greater entrepreneurial activity. Sudan-born Mo Ibrahim, an internationally known technology leader, established the telecommunications corporation Celtel. Ibrahim was not single-handedly responsible for the rapid increase of cell phone use across Africa, but he certainly played a large role in it. Since selling Celtel, his Mo Ibrahim Foundation has fought for good governance across the continent. In honor of their good governance accomplishments, the foundation has awarded the Ibrahim Prize for Achievement in African Leadership to Joaquim Chissano of Mozambique (2007), Festus Mogae of Botswana (2008), and Pedro Pires of Cape Verde (2011).

One can certainly argue that many Africans still live in desperate poverty, but many more live in conditions much improved since the 1970s and 1980s. The Mo Ibrahim Foundation publishes an "Ibrahim Index" to track progress toward improved economic and political conditions that can lead to expanded human-development opportunities. Kenya's Wangari Maathai won the 2004 Nobel Peace Prize for her "green belt movement" and its sustained commitment to environmental conservation. Even South Africa's successful hosting of the World Cup in 2010 bodes well for capacity building and management excellence. In short, the African future seems brighter than critics suggest.

     Christopher A. Whann

1. Iliffe, *Africans*, 1.

SOURCES

Bates, Robert. *When Things Fell Apart: State Failure in Late Century Africa.* New York: Cambridge University Press, 2008.

Collier, Paul. *The Bottom Billion: Why the Poorest Countries Are Failing and What Can Be Done About It.* New York: Oxford University Press, 2007.

Harms, Robert. *The Diligent: A Voyage Through the Worlds of the Slave Trade.* New York: Basic Books, 2003.

Herbst, Jeffrey. *States and Power in Africa: Comparative Lessons in Authority and Control.* Princeton, NJ: Princeton University Press, 2000.

Hochschild, Adam. *King Leopold's Ghost: A Story of Greed, Terror and Heroism in Colonial Africa.* New York: Mariner Books, 1999.

Hyden, Goran. *African Politics in Comparative Perspective.* New York: Cambridge University Press, 2006.

Iliffe, John. *Africans: The History of a Continent.* New York: Cambridge University Press, 2007.

Iliffe, John. *The African AIDS Epidemic: A History.* Athens: Ohio University Press, 2006.

Leonard, David, and Scott Straus. *Africa's Stalled Development: International Causes and Cures.* Boulder, CO: Lynne Rienner Publishers, 2003.

Mahajan, Vijay. *Africa Rising: How 900 Million African Consumers Offer More Than You Think.* Upper Saddle River, NJ: Wharton School Publishing, 2009.

Mo Ibrahim Foundation. http://www.moibrahimfoundation.org.

Ndulu, Benno. *Challenges of African Economic Development: Opportunities, Constraints, and Strategic Directions.* Washington, DC: International Bank for Reconstruction and Development/ The World Bank, 2007.

Scott, James. *Seeing Like a State: How Certain Schemes to Improve the Human Condition Have Failed.* New Haven, CT: Yale University Press, 1999.

Segal, Ronald. *Islam's Black Slaves: The Other Black Diaspora.* New York: Farrar, Strauss and Giroux, 2002.

U. S. Energy Information Administration. *Crude Oil and Total Petroleum Imports Top 15 Countries.* http://www.eia.doe.gov/pub/oil_gas/petroleum/data_publications/company_level_imports/current/import.html.

Young, Crawford. *The African Colonial State in Comparative Perspective.* New Haven, CT: Yale University Press, 1994.

EL ANATSUI
Bukpa Old Town, 2009
Aluminum and
copper wire
120 × 84 inches

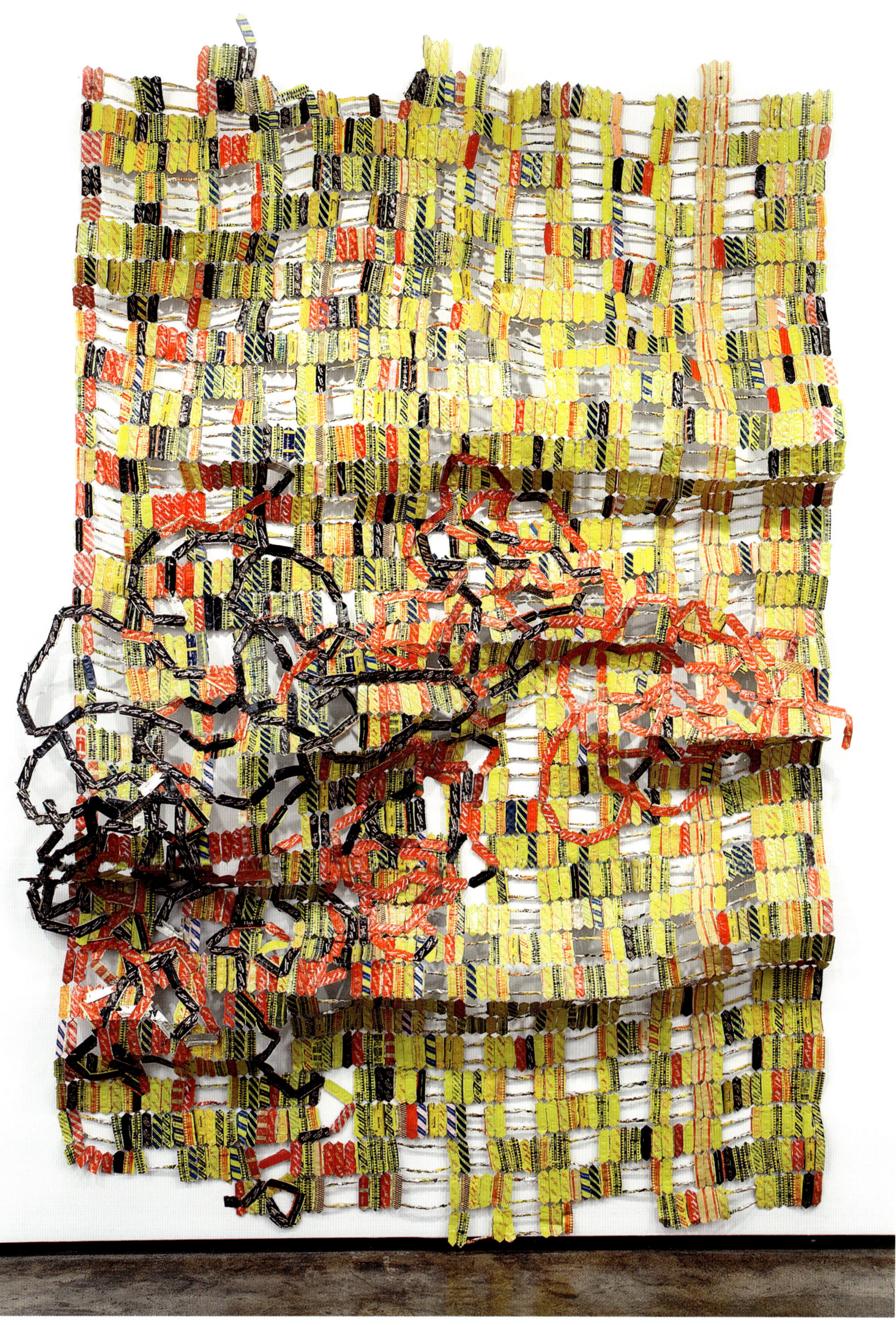

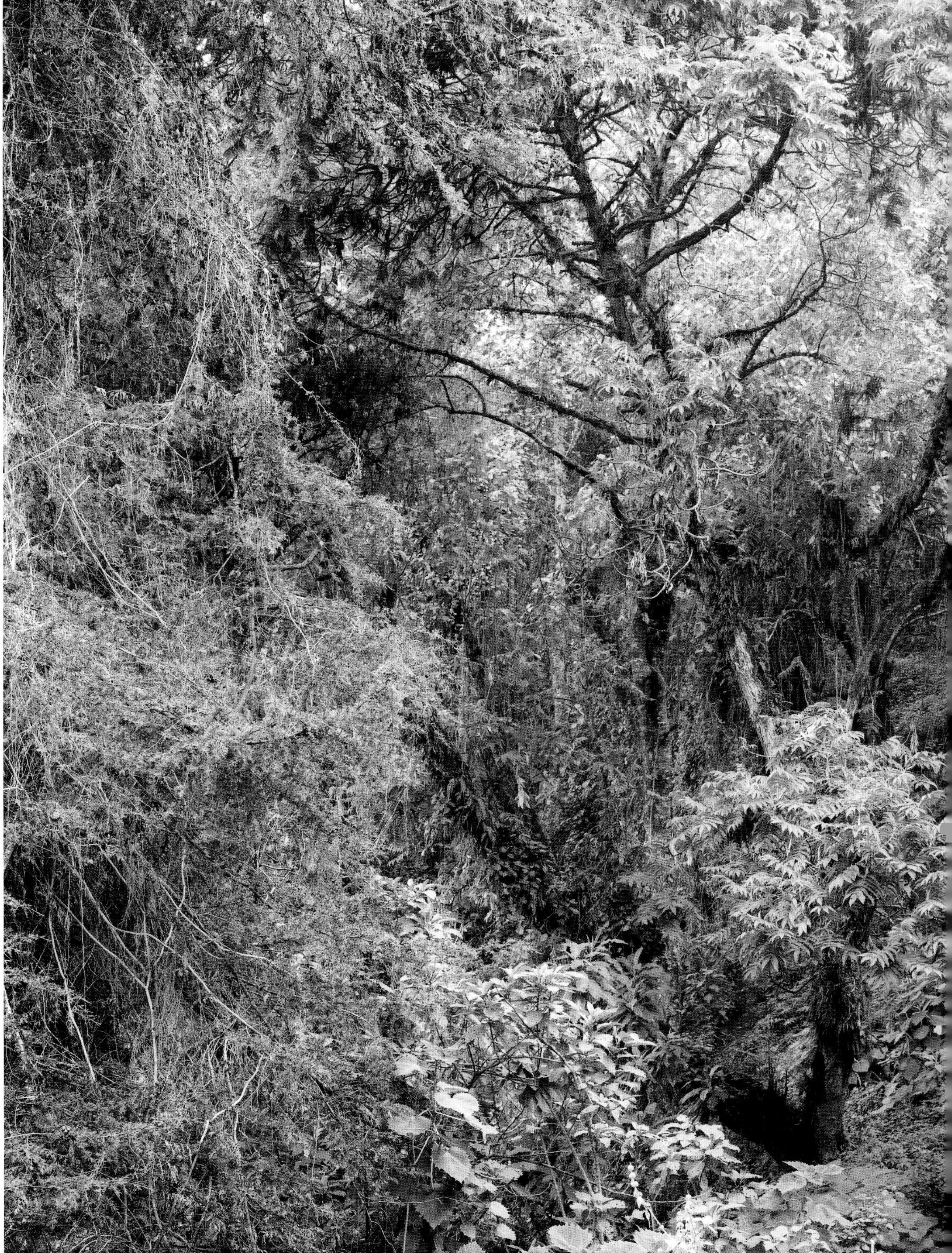

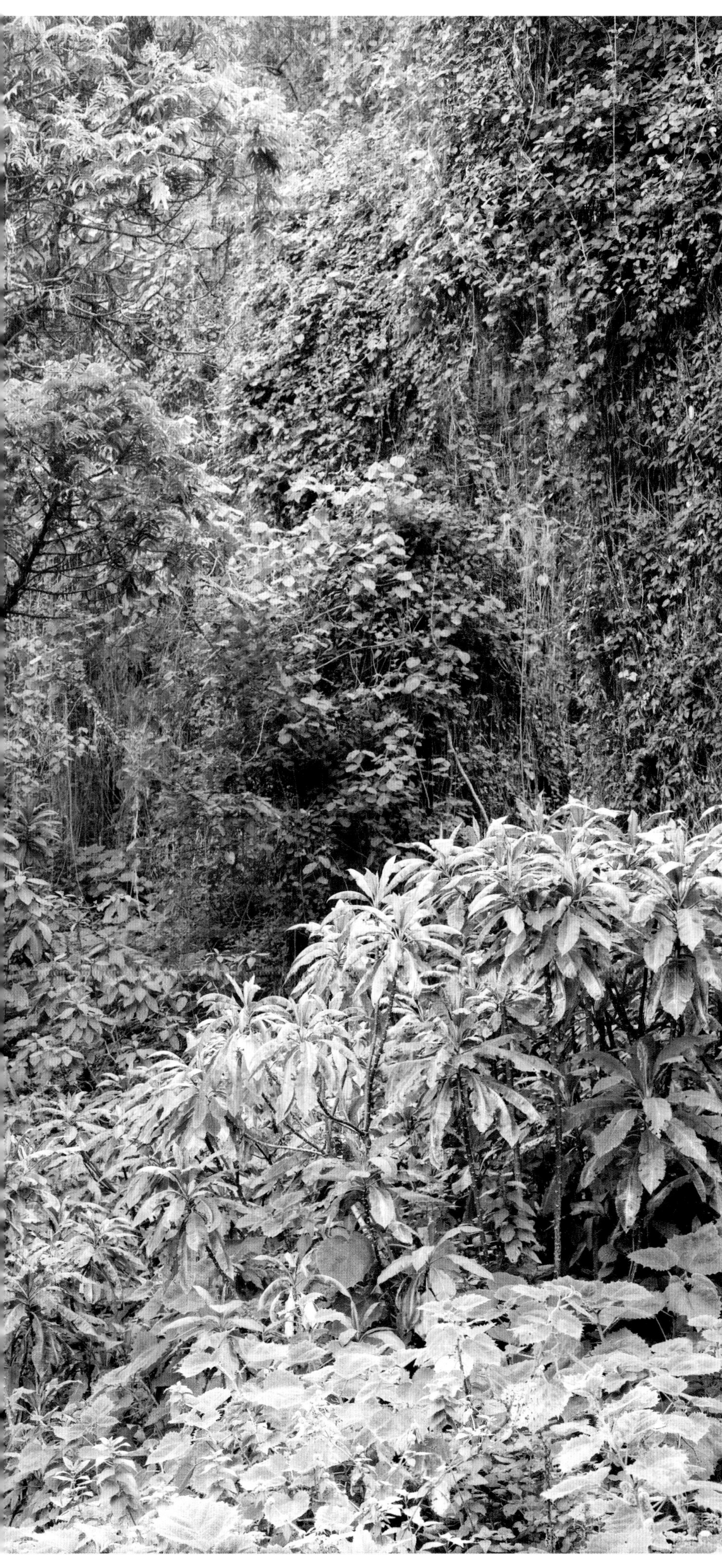

GARTH MEYER
Untitled (Figure #26), 2010
Rwanda
Silver halide hand print
10¼ × 13⅛ inches

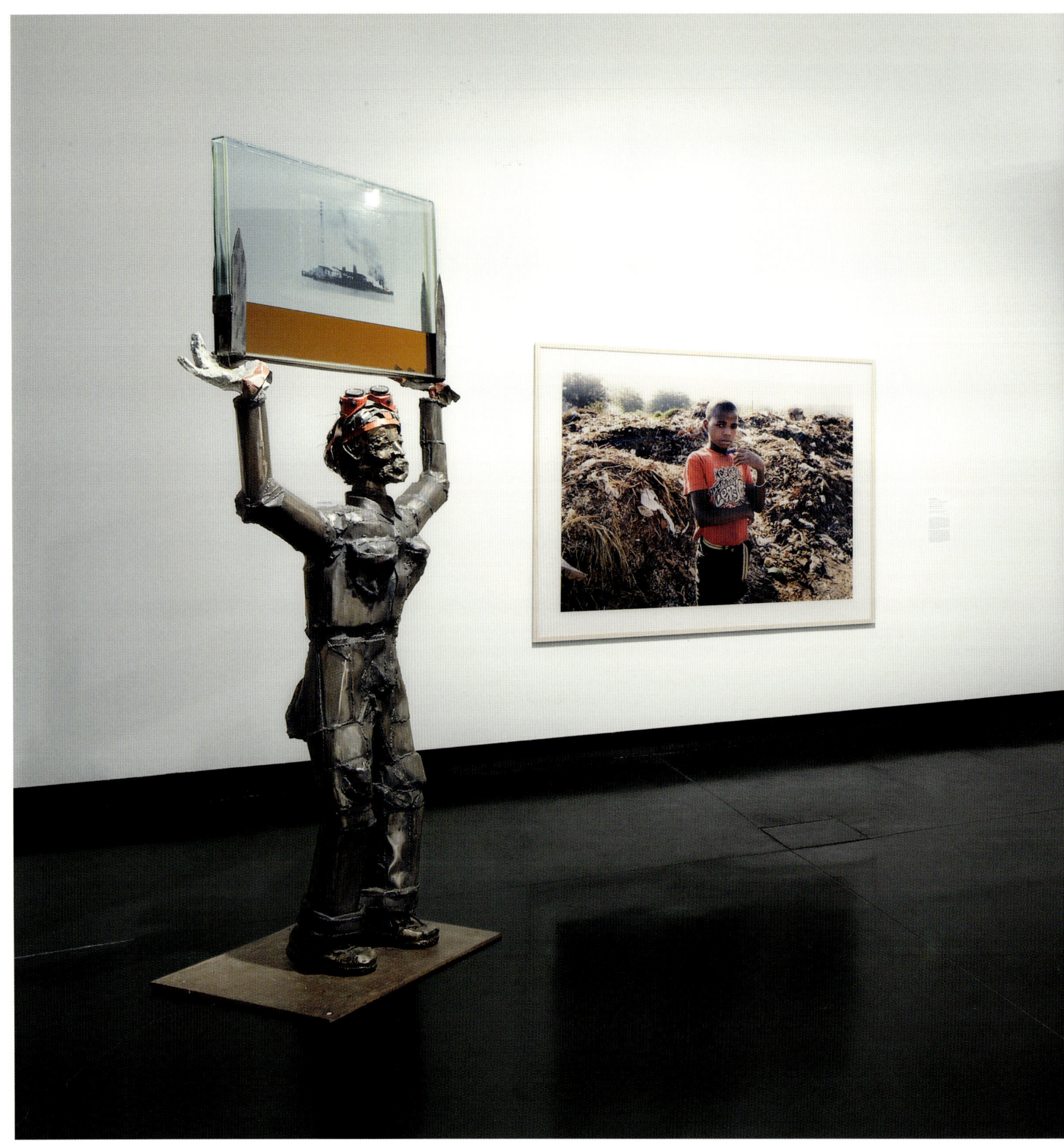

SOKARI DOUGLAS CAMP
AND ZWELETHU MTHETHWA
Installation view,
*Environment and Object ·
Recent African Art*,
Tang Museum, 2011

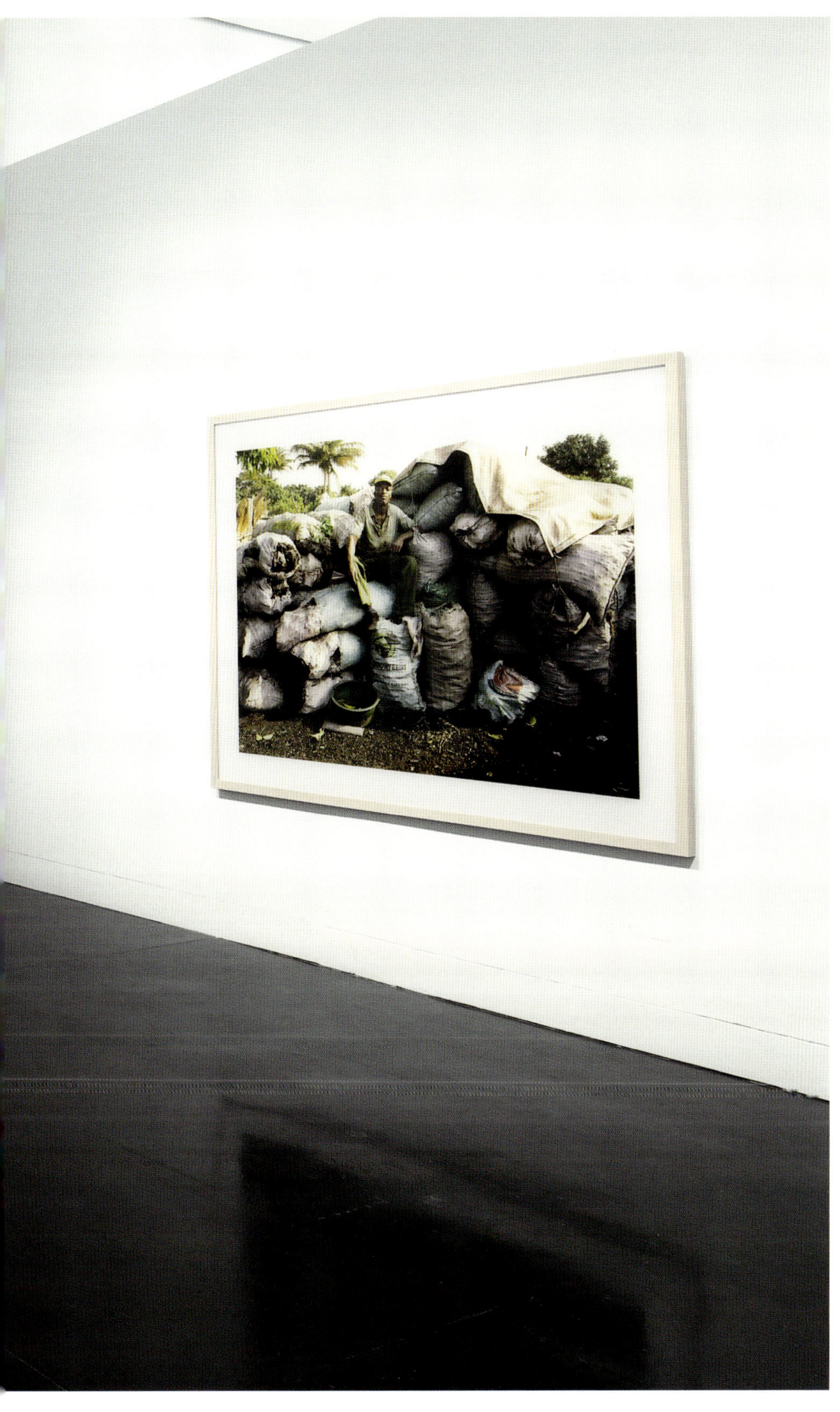

FOLLOWING PAGE
BARTHÉLÉMY TOGUO
Stupid African
President 1, 2006
Digital inkjet print
42⅞ × 30⅞ inches

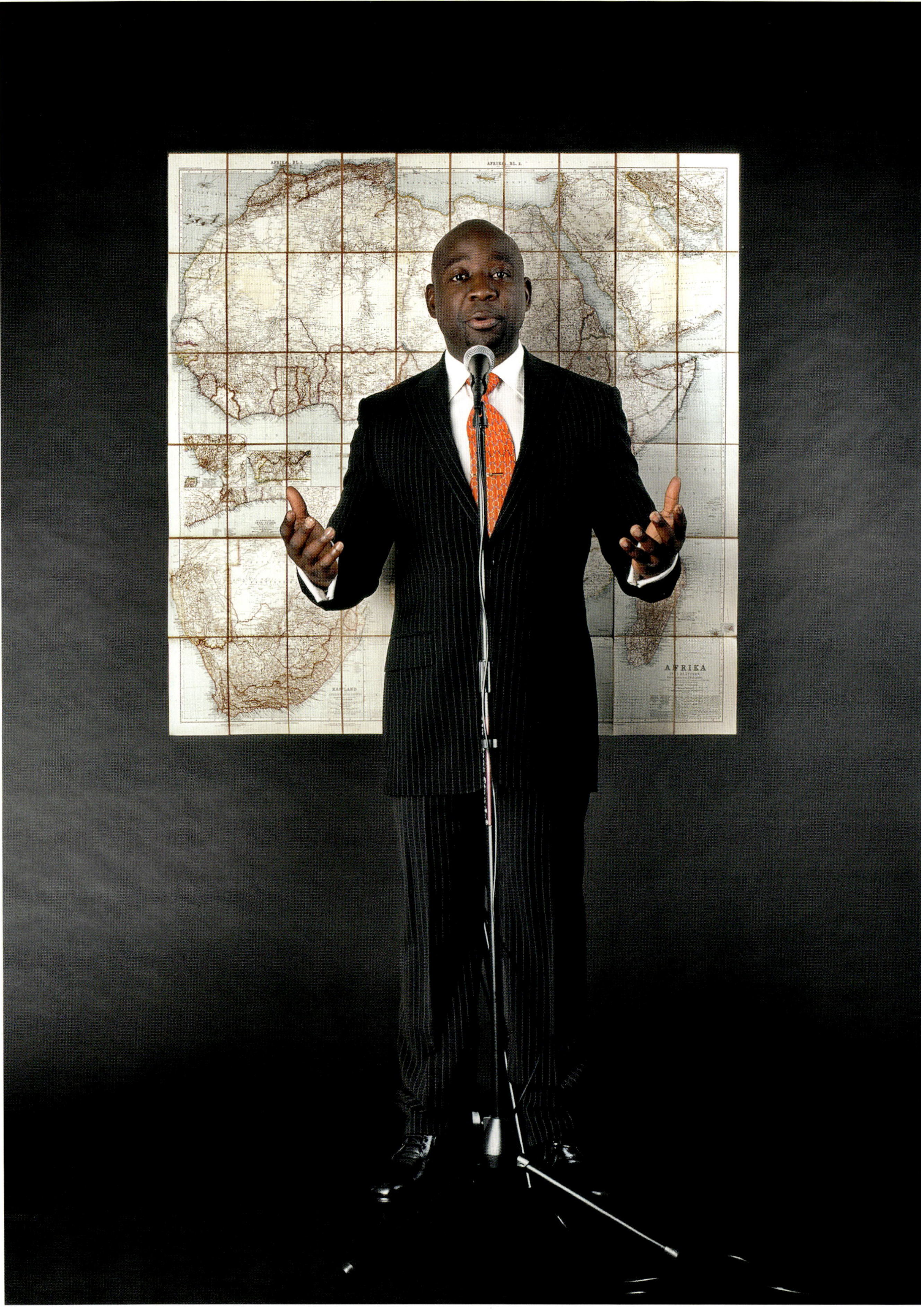
AFRIKA, Bl. 1
AFRIKA, Bl. 2
AFRIKA
KAPLAND

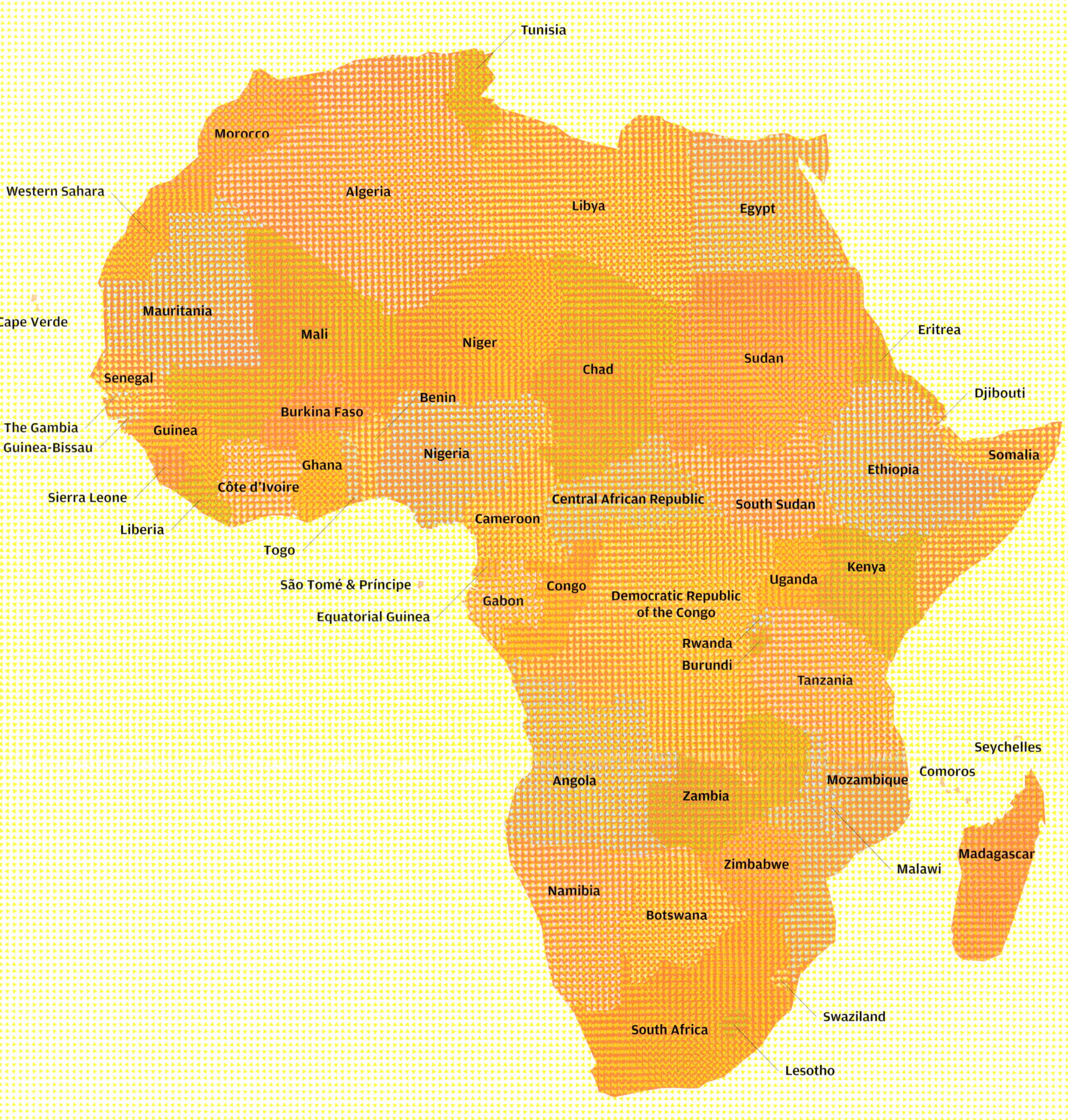
Tunisia
Morocco
Western Sahara
Algeria
Libya
Egypt
Cape Verde
Mauritania
Mali
Niger
Chad
Sudan
Eritrea
Senegal
Djibouti
The Gambia
Guinea-Bissau
Guinea
Burkina Faso
Benin
Nigeria
Somalia
Ethiopia
Sierra Leone
Ghana
Côte d'Ivoire
Central African Republic
South Sudan
Liberia
Cameroon
Togo
São Tomé & Príncipe
Gabon
Congo
Democratic Republic
of the Congo
Uganda
Kenya
Equatorial Guinea
Rwanda
Burundi
Tanzania
Seychelles
Comoros
Angola
Zambia
Mozambique
Zimbabwe
Malawi
Madagascar
Namibia
Botswana
Swaziland
South Africa
Lesotho

Lisa Aronson is an Associate Professor in the Department of Art History at Skidmore College. Aronson received her B.A. from Wayne State University, and her M.A. and Ph.D. from Indiana University. A scholar of African art history, Aronson has done extensive field research in Nigeria, the Ivory Coast, and Ghana. Her writing and research focuses on the impact of trade and colonialism on African textile production, issues of gender in African art, and contemporary African art, including photography. Supported by a Getty Collaborative Research Grant, she is currently coauthoring a book on the late nineteenth century Nigerian photographer Jonathan Adagogo Green.

Mark Auslander is an Associate Professor of Anthropology and Director of the Museum of Culture and Environment at Central Washington University. He is also currently a senior fellow at the National Museum of African Art, Smithsonian Institution. Auslander received his M.A. and Ph.D. in anthropology from the University of Chicago. A sociocultural anthropologist, his research focuses on political and symbolic processes in Africa and the African Diaspora. He has curated several exhibitions with students and community members of African and African American art, as well as of family and cultural history. He consulted on the Smithsonian Natural History Museum's permanent exhibition *African Voices* and worked on *Without Sanctuary: Lynching Photography in America* at the Martin Luther King, Jr., National Historic Site, Atlanta. His most recent publication is *The Accidental Slave Owner: Revisiting a Myth of Race and Finding an American Family* (University of Georgia Press, 2011).

Karen Kellogg is an Associate Professor of the Environmental Studies Program at Skidmore College, where she also directs the Water Resources Initiative. Kellogg studied industrial engineering at the University of Iowa and received her Ph.D. in ecology with conservation biology from Pennsylvania State University. Kellogg works with students on an interdisciplinary array of research topics, including historical water conflicts, the cultural significance of various fish species, the evolution and ecology of diverse fish communities, and renewable energy systems. Her research has been published in *Trends in Ecology and Evolution*, *Journal of Fish Biology*, and *African Journal of Ecology*, among others.

Chika Okeke-Agulu, an artist, curator, and art historian, is Assistant Professor in the Department of Art and Archeology and the Center for African American Studies at Princeton University. He was a fellow of the Clark Institute and Visiting Sterling Clark Professor at Williams College. Okeke-Agulu has taught at Pennsylvania State University, Emory University, University of Nigeria, Nsukka, and Yaba College of Technology, Lagos. He received his B.A. and M.F.A. from the University of Nigeria, Nsukka; M.A. in art history from the University of South Florida; and Ph.D. in art history from Emory University. Okeke-Agulu's curatorial projects include the Nigerian Section at the first Johannesburg Biennale (1995); *Seven Stories About Modern Art in Africa* at the Whitechapel Art Gallery, London (1995); *The Short Century: Independence and Liberation Movements in Africa, 1945–1994* at Museum Villa Stuck, Munich (2001); and *Who Knows Tomorrow* at the Nationalgalerie, Berlin (2010). He is coeditor of *Nka: Journal of Contemporary African Art* and coauthor (with Okwui Enwezor) of *Contemporary African Art Since 1980* (Damiani, 2009).

John S. Weber is the Dayton Director of the Frances Young Tang Teaching Museum and Art Gallery and Professor of Liberal Studies at Skidmore College. Previously, Weber worked as the Curator of Education and Public Programs at the San Francisco Museum of Modern Art and Curator of Contemporary Art at the Portland Art Museum in Oregon. He has taught at the San Francisco Art Institute, Mills College, the University of Washington, and the Pacific Northwest College of Art. Weber received his B.A. from Reed College and his M.F.A. from the University of California at San Diego. His exhibitions at the Tang Museum include *And Therefore I Am; Joachim Schmid: Photoworks 1982–2007*; and *Molecules That Matter*.

Christopher A. Whann is an Associate Professor of Business, Management, and Economics at Empire State College, State University of New York. Whann earned his B.A. in international relations and M.A. in political science at the University of Delaware, and his Ph.D. in political science and African studies at the University of Wisconsin–Madison. As a political scientist, Whann specializes in international business, emerging markets, and African and Caribbean business.

Dimensions listed by
height × width × depth
*Tour venues only

El Anatsui
Some Still Come Back, 2005
Aluminum and copper wire
92$\frac{1}{2}$ × 119 × 6$\frac{1}{2}$ inches
Mott-Warsh Collection,
Flint, Michigan
pages 95, 96

Bukpa New Town, 2009*
Aluminum and copper wire
115 × 75 inches
Courtesy of the artist and Jack
Shainman Gallery, New York

Bukpa Old Town, 2009*
Aluminum and copper wire
120 × 84 inches
Courtesy of the artist and Jack
Shainman Gallery, New York
page 137

Lara Baladi
Hope, 2010
Artist's book, photographs, and
CD with *Donkey Symphony*
7$\frac{7}{8}$ × 5$\frac{7}{8}$ inches
Courtesy of the artist, in
collaboration with Tang
Museum and Rautenstrauch-
Joest-Museum Kulturen
der Welt, Cologne
pages 56–59

Sammy Baloji
Untitled 13, 2006
Archival digital photograph on
satin matte paper
22$\frac{1}{2}$ × 93$\frac{1}{4}$ inches
Courtesy of the artist
and Axis Gallery, New York
pages 124–125

Untitled 21, 2006
Archival digital photograph on
satin matte paper
22$\frac{1}{2}$ × 63$\frac{1}{4}$ inches
Courtesy of the artist
and Axis Gallery, New York
pages 54–55

Jerry Buhari
*Images of Zaria City (Exploring
Images of a City)*, 2005
Watercolor on paper
7$\frac{1}{2}$ × 10$\frac{1}{2}$ inches
Courtesy of the artist
page 30

In the Pipeline, 2005
Acrylic on canvas
61$\frac{7}{8}$ × 54$\frac{1}{4}$ inches
Courtesy of the artist
page 27

Decay and Regeneration III, 2009
Termite-eaten wood, scrap
rubber, and acrylic
8$\frac{7}{8}$ × 24$\frac{7}{8}$ × 8$\frac{1}{2}$ inches
Courtesy of the artist
pages 28–29

Sokari Douglas Camp
Assessment, 1998
Steel, glass, acetate, and oil
95$\frac{1}{4}$ × 44 × 27$\frac{1}{8}$ inches
Zang Collection, London
page 51

Teasing Suicide, 2004
Steel and wood
73$\frac{1}{2}$ × 30 × 41 inches
Birgit Froese-Kindermann,
Düsseldorf
pages 52, 53

Viyé Diba
*Nous sommes nombreux, et
nos problèmes avec …* [We
are numerous, and our problems
with that …], 2008
Mixed media
Installation dimensions variable
Courtesy of the artist
pages 112–113, 114, 115

Parchemin de la Consommation
[Parchment of domestic
consumption], 2008
Mixed media
67 × 66$\frac{1}{4}$ × 7$\frac{7}{8}$ inches
Courtesy of the artist
page 116

Bright Ugochukwu Eke
Ripples and Storm 1, 2011
Recycled bottles and wire
Installation dimensions variable
Courtesy of the artist
pages 4–6, 99, 100–101, 102

Ripples and Storm 2, 2011
Salvaged wood and hardware
Installation dimensions variable
Courtesy of the artist
pages 100–101, 102

Romuald Hazoumé
Dan, 1992
Found objects, plastic, metal,
and rubber
15 × 16½ × 6 inches
CAAC – The Pigozzi Collection,
Geneva
page 111

Kaleta, 1992
Found objects, plastic, seeds,
cowries, rubber, and metal
21 × 14 × 7½ inches
CAAC – The Pigozzi Collection,
Geneva
page 122

La Reine [The Queen], 1995
Found objects, plastic, metal,
and acrylic
9 × 7¼ × 7½ inches
CAAC – The Pigozzi Collection,
Geneva
page 110

Gris-Gris, 2004
Found objects, plastic, and
synthetic hair
14½ × 10⅞ × 7 inches
CAAC – The Pigozzi Collection,
Geneva
page 123

Claudia Maigre, 2005
Found objects, plastic, and
synthetic and cotton thread
11½ × 10½ × 8 inches
CAAC – The Pigozzi Collection,
Geneva
page 122

Brut, 2001*
Found objects, metal, plastic, and
rubber
13½ × 14¾ × 3 inches
Courtesy of the artist and
October Gallery, London
page 122

Anna Fiata, 2002*
Found objects, plastic, metal, and
synthetic hair
18 × 12½ × 6½ inches
Courtesy of the artist and
October Gallery, London
page 122

Patatipatata, 2002*
Found objects, plastic, metal, and
synthetic hair
14½ × 8 × 6½ inches
Courtesy of the artist and
October Gallery, London
page 123

Tchin-Tchin BP!, 2010*
Found objects, plastic, and metal
90 × 27 × 27 inches
Courtesy of the artist and
October Gallery, London
pages 108–109

Garth Meyer
Untitled (Figure #1), 2004
Wild Coast, South Africa
Silver halide hand print
11⅜ × 14¼ inches
Courtesy of the artist,
Axis Gallery, New York, and
Rooke Gallery, Newtown,
Johannesburg
page 33

Untitled (Figure #4), 2009
Pangora, Gabon
Silver halide hand print
10¼ × 13⅛ inches
Courtesy of the artist,
Axis Gallery, New York, and
Rooke Gallery, Newtown,
Johannesburg
page 32

Untitled (Figure #12), 2000
Zimbabwe
Silver halide hand print
10¼ × 13⅛ inches
Courtesy of the artist,
Axis Gallery, New York, and
Rooke Gallery, Newtown,
Johannesburg
page 32

Untitled (Figure #26), 2010
Rwanda
Silver halide hand print
10¼ × 13⅛ inches
Courtesy of the artist,
Axis Gallery, New York, and
Rooke Gallery, Newtown,
Johannesburg
pages 32, 138–139

Zwelethu Mthethwa
Untitled (from the *Coal Miner
Series*), 2008
Chromogenic print
61½ × 79⅛ inches
Courtesy of the artist and Jack
Shainman Gallery, New York
pages 44–45

Untitled (from the *Contemporary
Gladiators Series*), 2008
Chromogenic print
60⅝ × 78¼ inches
Courtesy of the artist and Jack
Shainman Gallery, New York
pages 42–43

Nnenna Okore
Mbembe, 2009
Plastic bags
Dimensions variable,
approx. 137 × 153 × 110 inches
Courtesy of the artist
pages 104–105, 106

George Osodi
Blackened Explosion, 2005
Digital inkjet print
47½ × 31¾ inches
Courtesy of the artist and
Z Photographic, Surrey, England
page 41

Niger Boy, 2006
Digital inkjet print
47⅝ × 31⅞ inches
Courtesy of the artist and
Z Photographic, Surrey, England
page 40

*Oil Spill Near Farm Land
Ogoni*, 2007
Digital inkjet print
31¾ × 47½ inches
Courtesy of the artist and
Z Photographic, Surrey, England
page 38

Georgia Papageorge
*Africa Rifting: Lines of Fire
Namibia/Brazil*, 2001
Looped DVD, 15:39 minutes
Courtesy of the artist
page 127

Genesis, Crucifixion, Aftermath
(from *Africa Rifting*), 2003
Ultrachrome prints
3 parts, each 69⅜ × 47¼ inches
Courtesy of the artist
page 35–37, 127

Chéri Samba
Lutte contre les moustiques
[Fighting mosquitoes], 1989
Acrylic on canvas
25½ × 32 inches
Collection of Melva Bucksbaum
and Raymond Learsy
page 60

Yinka Shonibare, MBE
Black Gold I, 2006
Acrylic paint on Dutch wax-
printed cotton canvases
130 × 265 inches
National Museum of African Art,
Smithsonian Institution,
Museum Purchase, 2008-2-1
pages 46–47, 48–49

Black Gold II, 2006*
Acrylic paint on Dutch wax-
printed cotton canvases
130 × 265 inches
Zang Collection, London
page 47

Barthélémy Toguo
Stupid African President 1, 2006
Digital inkjet print
42⅞ × 30⅞ inches
Courtesy of the artist
page 142

Stupid African President 2, 2006
Digital inkjet print
42⅞ × 30⅞ inches
Courtesy of the artist
page 64

*Stupid African President 3
(Afrika Oil)*, 2006
Digital inkjet print
42⅞ × 30⅞ inches
Courtesy of the artist
page 63

Environment and Object · Recent African Art brings together art by African artists scattered across Africa, Europe, the United Kingdom, and the United States. It has proved a major undertaking, with works loaned from each of those countries and continents. As curators of the exhibition, we express our deepest appreciation to the many individuals and institutions whose cooperation, collaboration, loans, financial support, and excellent work made it possible.

First, we thank all of the artists for making their work available and supporting our conception for the exhibition. We are particularly grateful to Bright Ugochukwu Eke for spending a month on campus to create his new site-specific installation, *Ripples and Storm*, for the Tang, and to Viyé Diba for coming to Skidmore to install his piece, *Nous sommes nombreux, et nos problémes avec*. Many thanks as well to Barthélémy Toguo and George Osodi for speaking at the opening, and to Lara Baladi and Nnenna Okore for their visits to the Tang to talk about their work.

Financial support came from the Andrew W. Mellon Challenge Grant, which underwrites a range of collaborative efforts by Skidmore faculty and the Tang, including major faculty curatorial efforts such as this one. We also recognize with gratitude the support of Skidmore alumni Chief Oskar Ibru '81 and Chief Mrs. Wanda Ibru '79. Additional support was provided by the Tadahisu Kuroda Fund, the Virginia Gooch Puzak '44 Faculty Curatorial Endowment, *Institut Français*, the Creative Thought Fund at Skidmore, the Henry Luce Foundation, and Friends of the Tang. Bottles for Bright Eke's installation were donated by Springer Waste Management. We thank all of these donors for making *Environment and Object* possible.

We express our deep appreciation to catalogue essayists Mark Auslander of Central Washington University, Karen Kellogg of Skidmore College, and Christopher Whann of Empire State College, and to Chika Okeke-Agulu of Princeton University for his insightful interview. Their words were judiciously edited by Jay Rogoff, and the catalogue owes its splendid design to Barbara Glauber, Azusa Kobayashi, and Gregory Skiano of Heavy Meta. Thanks to Patricia Wieland for final edits and proofreading. Installation photography was done by Art Evans. We would also like to thank Ginny Puzak for her additional generous support of this catalogue.

We are also grateful to the aid and assistance of the following people at various stages of this project: Salah Hassan, Art Historian at Cornell University; Iris Berger, Historian, and Deborah LaFond, Librarian, at State University of New York, Albany; Curators

Chris Kreamer and Karen Milbourne and Librarian Janet Stanley at the Smithsonian National Museum of African Art; and William Karg, Director of the Contemporary African Art Gallery, New York.

We thank the many private and institutional lenders and partners whose generous cooperation made the exhibition possible, including

Lisa Brittan, Axis Gallery, New York
Melva Bucksbaum and Raymond Learsy
James Cohan Gallery, New York
Birgit Froese-Kindermann
Ziggi Golding, Z Photographic, Surrey, England
Elisabeth Lalouschek, October Gallery, London
Joachim Melchers, ARTCO Gallery, Herzogenrath, Germany
Mott-Warsh Collection, Flint, Michigan
National Museum of African Art, Smithsonian Institution, Washington, D.C.
Kerstin Pinther and the Rautenstrauch-Joest-Museum of Ethnography, Cologne
Gavin Rooke Gallery, Johannesburg
Jack Shainman Gallery, New York
Elisabeth Whitelaw, Pigozzi Collection, Geneva
James Zang, London

Environment and Object will travel to the Anderson Gallery of Virginia Commonwealth University and the Middlebury College Museum of Art, and we are grateful to colleagues Ashley Kistler of VCU and Emmie Donadio of Middlebury College for their interest and support of this project.

Many students were involved in the planning, research, and creation of the show. We thank Evan Cohen '12, James Eades '11, Rachel Fisher '12, and Melinda Kiefer '11, who worked with Bright Eke on his installation. David Serotte '11, Diane McGlauflin, and Arianna Galluzzo assisted in the preparation of the interactive online feature, and Jordan Klein '10 and Amanda Hawkins '12 assisted with exhibitions work. Many students in Lisa Aronson's art history classes assisted in exhibitions research, including Michelle Anderson '10, Matthew Awai '10, Rebecca Braungard '09, Ann Cascella '11, Gregory DelliCarpini '10, Rachel Downes '09, Francesca Fanelli '09, Allegra Gurfein '10, Katharine Hohorst '10, Dara Kruvant '10, Natalie Litton '10, Rachel Loeb '11, Michelle Lorenzutti '10, Abel Mills '11, John Tyler Norton '10, Emily Schenkein '09, Elizabeth Streeter '09, Hannah Kagan-Moore '12, Michaela Slavid '10, and Olive Wicherski '11.

In the Skidmore Geographic Information Systems Lab, Alex Chaucer and Andrew Noone '11 provided valuable assistance making maps for the gallery kiosks and online feature. We are grateful to Skidmore's Office of the Dean of Faculty for funding Noone and Lisa Aronson's 2010 Summer Faculty–Student Collaborative project, "Mapping and Visualizing the African Environment."

Environment and Object was ably installed by our installation crew, led by the Tang's Head of Installation, Chris Kobuskie, and our Senior Preparator, Torrance Fish. Crew members included Luke Anderson, Samuel Coe, Scott Heydrick, Sanford Mirling, Matthew Scellen, and Cynthia Zellner.

Throughout the Tang and Skidmore, many other staff members have contributed to the show's success. We salute Curatorial Assistants Kristen Carbone Boyle and Megan Hyde, both for their curatorial collaboration and for keeping myriad exhibition details moving forward on schedule. The Tang's Chief Curator and Associate Director, Ian Berry, offered sage curatorial advice and logistical support in many areas. Designer Patrick O'Rourke provided excellent exhibition cards, posters, and wall graphics. Head Registrar Elizabeth Karp brilliantly strategized and oversaw a complex schedule of shipments from three continents. Assistant Registrar Ryan Lynch provided skilled conditioning. Digital Resources Content Producer Vickie Riley and Senior Educator Susi Kerr spearheaded the online interactive program and other education programs on- and offline, and Museum Educator Ginger Ertz contributed family, teacher, and school programming. In the Tang's administrative office, Kelly Ward and Lorri Riggs provided excellent travel, event, and logistical support, while Jennifer Napierski attended ably to financial matters. Barbara Schrade, the Tang Store and Community Relations Manager, ensured that the exhibition was supported through relevant publications and marketing. Robert Kimmerle, Skidmore's Director of Community Relations ably handled public and press relations.

Finally, we thank our many faculty colleagues at Skidmore who incorporated the exhibition into their teaching, and we express our profound gratitude to the Skidmore community of trustees and alumni, as well as Skidmore President Philip A. Glotzbach, Vice President for Academic Affairs Susan Kress, and Dean of the Faculty Muriel Poston for their support of the Tang Museum and such ambitious projects as *Environment and Object · Recent African Art*.

Lisa Aronson, *Associate Professor of Art History*
John S. Weber, *Dayton Director, The Frances Young Tang Teaching Museum and Art Gallery*

Skidmore College, Saratoga Springs, New York